ESP 实用视听说系列教材

【总主编 赵团结】

机械工程英语

主　编：陈思雯

副主编：李　娥　王艳荣

审　校：堵海鹰　赵团结

Mechanic
Engineering
English

首都经济贸易大学出版社

Capital University of Economics and Business Press

·北 京·

图书在版编目(CIP)数据

机械工程英语/陈思雯主编. --北京:首都经济贸易大学
出版社,2020.2
ISBN 978-7-5638-3060-2

I.①机… II.①陈… III.①机械工程—英语—高等学校—
教材 IV.①TH

中国版本图书馆 CIP 数据核字(2020)第 009251 号

机械工程英语
主编 陈思雯
副主编 李 娥 王艳荣
Jixie Gongcheng Yingyu

责任编辑 刘元春
封面设计 风得信·阿东
         FondesyDesign
出版发行 首都经济贸易大学出版社
地   址 北京市朝阳区红庙 (邮编 100026)
电   话 (010)65976483 65065761 65071505(传真)
网   址 http://www.sjmcb.com
E-mail   publish@ cueb.edu.cn
经   销 全国新华书店
照   排 北京砚祥志远激光照排技术有限公司
印   刷 北京玺诚印务有限公司
开   本 787 毫米×1092 毫米 1/16
字   数 346 千字
印   张 13.5
版   次 2020 年 2 月第 1 版 2020 年 2 月第 1 次印刷
书   号 ISBN 978-7-5638-3060-2
定   价 38.00 元

序 言
PREFACE

改革开放以后，为加强对外交往，英语被国家教育部门确定为大学必修课程。从此，万千中国学子投入大量的时间与精力去学习英语。这种长时间的重视与投入，极大地提高了国民的英语水平，密切了中外交往，加快了中国现代化进程，从而有力地促进了中国特色社会主义伟大事业的发展。经历了逾40年改革开放风雨洗礼的中国，正逐步走上世界舞台的中央。越来越有中国特色社会主义道路自信、理论自信、制度自信、文化自信的中国人民开始重新审视这种不分专业、齐头并进的全民大学英语教育，在客观、理性的探讨中已逐步形成共识，那就是：大学英语教育必须改革。可是，该如何改革呢？

回望过去是为了规划未来！中华儿女学习英语的热情在学习西方先进科技、开放图强的怒吼中被点燃，在追赶西方、复兴文化的脚步声中日渐高涨。时至今日，我们的英语学习怎能忘却初心、迷失方向，甚或不自觉地成为自毁文化长城的西方特洛伊木马？我们的大学英语教育改革必须把正航向。

知易行难！"雄关漫道真如铁，而今迈步从头越！"武昌工学院是以培养具有创新精神的应用型人才为己任的工科院校，必须以舍我其谁的气魄，在大学英语教学改革中勇当开路先锋。为此，对武昌工学院的大学英语教学改革，我提出了落实"两个一千"（1 000个专业词汇，1 000个专业句子）的具体目标。"功崇惟志，业广惟勤"，武昌工学院全体英语教师，在赵团结同志的带领下，志存高远，攻坚克难，积极推进ESP（English for Specific Purposes）大学英语教学改革，花费大量心血，在深入调查研究的基础上，精心编写了这套"ESP实用视听说系列教材"。"德不孤，必有邻"，我相信，这套涵盖七大专业门类的英语教材必能引起同行的关注，得到他们的爱护、响应与支持。"大厦之成，非一木之材也；大海之阔，非一流之归也"，希望本套英语教材的全体编写同志，戒骄戒躁，汇智聚力，与兄弟院校的同行携手并进，不断提高教材质量，悉心将它打造成业界精品，共同回应时代要求，开辟大学英语教学改革的新天地！

无尽今来古往，多少春花秋月！任何事业都是在后人赶超前人的奋斗中，不断自我更新扬弃，从而长盛不衰的。"芳林新叶催陈叶，流水前波让后波"，英语教学改革永远在路上。希望同志们不断赶超前人，持续推进英语教学改革。"合抱之木，生于毫末；九层之台，起于累土。"ESP大学英语教学改革，已经迈出了坚实的第一步，必能善始善终，行稳致远，收前人未竟之功。

是为序。

武昌工学院董事长兼校长　李勇

2019年12月15日

本书使用说明

INSTRUCTIONS

　　"ESP 实用视听说系列教材"是武昌工学院国际教育学院组织编写的一套教材,适用于普通高等院校本科专业大三的学生。该系列教材以学生专业学习阶段的专业词汇和专业句子为基础,以视听和口语练习为主要教学内容,强化学生的专业英语应用能力。本系列教材共 7 个分册,分别是:《机械工程英语》《土木工程英语》《信息工程英语》《食品化工英语》《艺术设计英语》《经济管理英语》《财务会计英语》。每个分册包括 10 个单元,覆盖相关专业领域内的主要专业或者方向。选用这套教材的学校,可以根据学生的专业灵活调整单元的顺序,自由选择教学的内容。

　　每个单元包括六个模块,分别是:①Pre-class Activity;②Specialized Terms;③Watching and Listening;④Talking;⑤After-class Exercises;⑥Additional Reading。第一个模块是课前活动,为学生简要介绍相关领域里的名人,让学生预习,并进行自由讨论,作为课前热身。第二个模块要求学生课前预习专业词汇,教师可以在课堂上采用听写、词语接龙等形式检查学生对专业词汇的记忆效果。第三个模块是视听训练,所选视频均来自国外主流网站最新的视频资料,并根据学生的接受程度,根据由易到难的原则,编写了选择题、判断题、填空题和讨论题。第四个模块是口语训练,包括需要学生诵读的 100 个经典句子(50 个通用句子和 50 个专业句子),以及供班级组织活动的两个对话样板,然后进行模仿练习和较高难度的讨论和辩论练习。第五个模块主要是结合本单元视听材料中出现的专业词汇,编写了配对、填空、翻译、写作等相关应用题型。第六个模块是课后阅读,主要内容是相关领域内的世界著名公司的简介,并采用四级考试中的长篇阅读题型,加强学生信息捕获的能力。

　　每个单元大约需要 6 个学时。建议第一、二模块用 0.5 个学时,第三模块大概需要 2.5 个学时,第四模块需要 2 个学时,第五、六模块需要 1 个学时。为了方便师生使用本教材,我们把视频文字和练习答案,放在了出版社的网络平台上,扫描书中的二维码,即可查阅;我们还提供了相关听力视频的网址供广大师生查阅。如果无法在网上查到视频或有任何疑问和批评建议,可以联系本套丛书的总主编赵团结教授,他的联系方式如下:

手机:13986141410

座机:027-88142008

邮箱:markztj@ sina.com

QQ 号:3153144383

微信号:13986141410

编者

2020 年 1 月 3 日

CONTENTS

Unit One   Mechanical Design

I. Pre-class Activity

Directions: *Please read the general introduction about* **James Watt** *and tell something more about the great scientist to your classmates.*

James Watt(1736—1819) was a Scottish inventor, mechanical engineer, and chemist, who improved on Thomas Newcomen's 1712 Newcomen steam engine with his Watt steam engine in 1781, which was fundamental to the changes brought by The Industrial Revolution in both his native Great Britain and the rest of the world.

While working as an instrument maker at the University of Glasgow, Watt became interested in the technology of steam engines. He realized that contemporary engine designs wasted a great deal of energy by repeatedly cooling and reheating the cylinder. Watt introduced a design enhancement, the separate condenser, which avoided this waste of energy and radically improved the power, efficiency, and cost-effectiveness of steam engines. Eventually he adapted his engine to produce rotary motion, greatly broadening its use beyond pumping water.

Watt attempted to commercialize his invention, but experienced great financial difficulties until he entered a partnership with Matthew Boulton in 1775. The new firm of Boulton and Watt was eventually highly successful and Watt became a wealthy man. In his retirement, Watt continued to develop new inventions though none was as significant as his steam engine work. He died in 1819 at the age of 83. He developed the concept of horsepower, and the SI unit of power, the watt, was named after him.

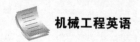

II. Specialized Terms

Directions：*Please memorize the following specialized terms before class so that you will be able to better cope with the coming tasks.*

actuating motor 伺服电机

arrester brake 制动器

assembling n. 装配

benchmark n. 基准

bore n. 钻削,镗削

boring machine 镗床

broaching machine 拉床

broaching n. 拉孔

chuck n. 卡盘

clearance angle 后角

coupling n. 联轴器

crosshead n. 十字结联轴节

cutting depth 切削深度

data storage[计]数据存储

deformation n. 变形

depth of hardening 硬化深度

depth recorder 深度记录器

derivative control 微分控制

derived unit 导出单位

derivable adj. 可诱导的;
可推论的;可引出的

design n. 设计;图案 vt. 设计;
计划;构思

design drawing 设计图

design failure 设计失败

design height 设计高度

desktop n. 桌面;台式机

deviation n. 偏差,偏移,背离

device n. 器件

diameter n. 直径

digital signal 数字信号

diode n.[电子]二极管

dioxide n. 二氧化物

discipline n. 学科;纪律;训练;惩罚 vt. 训
练,训导;惩戒

discrimination n. 识别力

disk space 磁盘空间

displace vt. 取代,替代

displacement n. 位移

display screen 显示器,显示屏

distal adj. (肌肉、骨、肢体等)末端的

distillation n. 精馏,蒸馏,净化;蒸馏法;精
华,蒸馏物

drill machine 钻床

durability n. 耐用度

finish machining 精加工

finite element 有限元

fluid clutch 液压驱动泵

forge n. 锻

found n. 铸造

gear n. 齿轮

gear machining 齿轮加工

gear shaping 插齿

gearbox casing 变速箱体

grinder n. 磨床

headstock n. 主轴箱

helix n. 螺旋

hobbing n. 滚齿

hydraulic pressure 液压

hydraulic pump 液压泵

intensity n. 强度

internal force 内力

invalidation n. 失效

lathe tool 车刀

lead rail 导轨

load n. 载荷

locksmith n. 钳工

machining center 加工中心

medium n. 介质

mill n. 铣削

milling cutter 铣刀

milling machine 铣床

nose of tool 刀尖

oxidation n. 氧化

pin n. 销

planing n. 龙门刨削

power n. 功率

rake angle 前角

rake face 前刀面

reliability n. 可靠性

rigidity n. 刚度

rolling bearing 滚动轴承

rough machining 粗加工

rust v. 腐蚀

safety factor 安全系数

screw rod 丝杠

section n. 截面

sliding bearing 滑动轴承

spindle n. 主轴

spline n. 键

spring n. 弹簧

stability n. 稳定性

stamping n. 压模

stepper motor 步进电机

strap n. 皮带

stress n. 应力

subassembly n. 组件

tangent n. 切线

thread n. 螺纹

turning n. 车削

wear v. 磨损

weld n. 焊

workpiece n. 工件

III. Watching and Listening

Task One   Day in the Life of a Mechanical Engineer

视频链接及文本

New Words

mechanical adj. 机械的

alloy n. 合金

constraint n. 限制

prototype n. 原型

application n. 应用

iterate vt. 重复,迭代

finalize vt. 完成

tinker vi. 焊补

internship n. 实习

hierarchy n. 等级制度

consultancy n. 顾问,咨询

portfolio n. 公文包

Exercises

1. *Watch the video for the first time and choose the best answer to the following questions.*

   1) What is the job of Nivay according to the video? _____

      A. accountant

      B. mechanical engineer

      C. light engineer

      D. physician

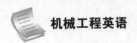

2) Nivay is working on the product of _____.

    A. model car　　　　　　　　B. model plane

    C. studio headphone　　　　　D. studio watch

3) The application that we use the most is _____.

    A. CAD　　　　　　　　　　B. PS

    C. CPU　　　　　　　　　　D. VPN

4) How much did Nivay know about mechanics in college? _____

    A. very well　　　　　　　　B. a lot

    C. nothing　　　　　　　　　D. not exactly

5) When did Nivay work for Alloy Product Development? _____

    A. in high school　　　　　　B. in college

    C. after graduation　　　　　D. after internship

2. *Watch the video again and decide whether the following statements are true or false.*

1) Nivay Anandarajah is a mechanical engineer for Alloy Product Development. (　)

2) He works with the 2D geometry and the design. (　)

3) The application that they use the most is CAD(computer-aided design). (　)

4) He has made thirteen different prototypes. (　)

5) He got an internship before he worked there years later. (　)

3. *Watch the video for the third time and fill in the following blanks.*

    In high school, you know, I liked to tinker with stuff you do some _____ and little bit of _____ here and there. Throughout college I didn't know exactly what in _____ engineering I wanted to do, but you know I found myself drawn towards _____ experience and started the _____ process because of the creativity. I did an _____ which I found was too large of a company and then I did an internship in a small design _____ which was two people and then this is sort of where I came after school, which was that the 20-person _____ where it's small enough where everyone has a say and it's a _____ hierarchy, so it doesn't feel like you're trying to climb a corporate ladder but at the same time there's lots of people that you can learn from and get _____ from.

4. *Share your opinions with your partners on the following topics for discussion.*

1) How do you like the day in the life of a mechanical engineer? Please summarize the features of a mechanical engineer.

2) Can you use a few lines to list what's your understanding about mechanical design? Please use an example to clarify your thoughts.

Task Two　2D Sketching

New Words

utilize vt. 利用,使用　　　　　　　　　　sketch n. 草图;素描

视频链接及文本

feature n. 特征,特点

geometry n. 几何形状

dimension n. 尺寸;维,维度

entity n. 实体

horizontal adj. 水平的

vertical adj. 垂直的

parallel adj. 平行的

perpendicular adj. 成直角的；直立的

concentric adj. 同轴的

modification n. 修改

Exercises

1. *Watch the video for the first time and choose the best answer to the following questions.*

   1）According to the video, Computer-aided design or CAD mainly involves _____.

   　　A. drawings 　　　　　　　　B. calculation

   　　C. movies 　　　　　　　　　D. internet

   2）The 3D features come into being with a third _____ on the basis of 2D sketch.

   　　A. triangle 　　　　　　　　B. circle

   　　C. line 　　　　　　　　　　D. dimension

   3）Which must be taken into account upon creation of the geometry in 2D programs? _____

   　　A. safety 　　　　　　　　　B. accuracy

   　　C. comfort 　　　　　　　　　D. convenience

   4）Which is not one of the common examples of sketch relationships? _____

   　　A. The sketches are horizontal. 　　B. The sketches are vertical.

   　　C. The sketches are curved. 　　　　D. The sketches are parallel.

   5）Which of the following is true about the 3D CAD applications? _____

   　　A. There are no other substitutes in terms of drawing design.

   　　B. They allow for the generation of 2D sketch relationships.

   　　C. There are more than twenty 3D CAD applications in the market.

   　　D. They allow for future addition of 4D elements.

2. *Watch the video again and decide whether the following statements are true or false.*

   1）Dimensions are used to reflect the location of the geometry. （　）

   2）In many 2D programs the dimensions drives the geometry. （　）

   3）The dimensions can change the sketch geometry. （　）

   4）The 2D relationships allow the designer to apply design intent at the sketch level. （　）

   5）This CAD applications don't allow access for future modifications. （　）

3. *Watch the video for the third time and fill in the following blanks.*

   　　Then dimensions are used to reflect the _____ of the geometry. In other words the geometry drives the _____. This is significantly different in most _____ CAD applications, because now the dimensions can change the _____ geometry. This allows for more rapid _____ and part _____ capabilities. Additionally many 3D CAD applications allow for the _____ of 2D sketch relationships. These relationships can be either _____ along with the sketch _____ or _____ afterwards.

4. *Share your opinions with your partners on the following topics for discussion.*

　　1）Do you know the features of 2D sketches? What about the 3D design?

　　2）Can you design a drawing with the aid of CAD?

IV. Talking

Task One　Classical Sentences

Directions：*In this section，some popular sentences are supplied for you to read and to memorize. Then，you are required to simulate and produce your own sentences with reference to the structure.*

General Sentences

1. Where do you live?
   你住在哪里？

2. I live on the Washington Street.
   我住在华盛顿街。

3. I'm Mr. Smith's next door neighbor.
   我是史密斯先生的邻居。

4. You live here in the city，don't you?
   你住在这个城市，对吗？

5. I'm from out of town.
   我住在城外。

6. How long have you lived here?
   你在这里住了多久了？

7. I've lived here for five years.
   我在这里住了 5 年了。

8. Where did you grow up?
   你在哪儿长大？

9. I grew up right here in this neighborhood.
   我就在这儿附近长大。

10. My friend spent his childhood in California.
    我朋友是在加利福尼亚度过他的童年的。

11. He lived in California until he was seventeen.
    他 17 岁以前都住在加利福尼亚。

12. There have been a lot of changes here in the last 20 years.
    在过去的 20 年间，这里变化很大。

13. There used to be a grocery store on the corner.

拐角处以前有一个杂货店。

14. All of those houses have been built in the last ten years.
　　那些房子都是最近十年里建成的。

15. They're building a new house up the street.
　　这个街上正在建一座新房子。

16. If you buy that house, will you spend the rest of your life there?
　　如果你买了那栋房子，你会在那里度过余生吗？

17. Are your neighbors friendly?
　　你的邻居们友好吗？

18. We all know each other pretty well.
　　我们都很了解对方。

19. Who bought that new house down the street?
　　大街那头的那栋新房子谁买下了？

20. An old man rented the big white house.
　　一个老人租了那栋白色大房子。

21. We're looking for a house to rent for the summer.
　　我们在找一栋房子夏天租住。

22. Are you trying to find a furnished house?
　　你是不是在找一栋带装修的房子？

23. That house is for sale. It has central heating. It's a bargain.
　　那栋房子在出售，中央供暖，价格很合理。

24. This is an interesting floor plan. Please show me the basement.
　　这是一个不错的平面图。麻烦带我到地下室去看看。

25. The roof has leaks in it, and the front steps need to be fixed.
　　屋顶漏水，前面台阶也需要修理。

26. We've got to get a bed and a dresser for the bedroom.
　　我们得在卧室弄张床和一个梳妆台。

27. They've already turned on the electricity. The house is ready.
　　房子已经通电，可以入住了。

28. I'm worried about the appearance of the floor.
　　我对地板的外观甚感担忧。

29. What kind of furniture do you have? Is it traditional?
　　你这里有什么样式的家具？是传统型的吗？

30. We have drapes for the living room, but we need kitchen curtains.
　　客厅的窗帘我们有了，但我们需要厨房的窗帘。

31. The house needs painting. It's in bad condition.
　　这房子得粉刷了，情况很糟。

32. In my opinion, the house isn't worth the price they're asking.

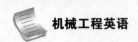

依我看,这间房子根本不值他们要的价钱。

33. Will you please measure this window to see how wide it is?
请你测量一下这个窗口,看看它有多宽?

34. This material feels soft.
这种材料摸上去很软。

35. —Can you tell me where Peach Street is?
—Two blocks straight ahead.
——你能告诉我桃园街在哪儿吗?
———直朝前走,过两个街区就到了。

36. Should I go this way, or that way?
我要走这条路还是那条?

37. Go that way for two blocks, and then turn left.
走那条路,穿过两个街区后向左拐。

38. How far is it to the university?
到大学还有多远?

39. The school is just around the corner. It's a long way from here.
学校就在拐角处。从这儿走,还有很长的一段路。

40. Are you married?
你结婚了吗?

41. No, I'm not married. I'm still single.
没有,我还是单身。

42. Your niece is engaged, isn't she?
你的侄女订婚了,是吗?

43. When is your grandparents' wedding anniversary?
你祖父母的结婚纪念日是什么时候?

44. How long have they been married?
他们结婚多久了?

45. They've been married for three decades.
他们结婚三十年了。

46. We're trying to plan our future.
我们在努力计划我们的未来。

47. I've definitely decided to go to California.
我已经决定去加利福尼亚了。

48. Who are you writing to?
你在给谁写信呢?

49. I'm writing to a friend of mine in South American.
我在给南美的一位朋友写信。

50. —How long has it been since you've heard from your uncle?

—I feel guilty because I haven't written to him lately.

——你多久没收到你叔叔的信了？

——我总觉得很内疚,因为我最近没有给他写信。

Specialized Sentences

1. Metals are elements that generally have good electrical and thermal conductivity.

   金属是具有良好导电性和导热性的元素。

2. Examples of metal alloys include stainless steel which is an alloy of iron, nickel, and chromium; and gold jewelry which usually contains an alloy of gold and nickel.

   金属合金的例子有不锈钢,它是一种铁、镍、铬的合金;金饰品即含有金、镍的合金。

3. Many metals and alloys have high densities and are used in applications which require a high mass-to-volume ratio.

   许多金属和合金具有高密度,因此被用在需要较高质量体积比的场合。

4. Belt drives offer a maximum of versatility as power transmission elements.

   皮带传动装置作为传递动力的元件用途极广。

5. They allow the designers have considerable flexibility in location of driver and driven machinery.

   这种装置使设计人员在确定主动机械与从动机械的位置时有相当大的灵活性。

6. Tolerances are not critical as is the case with gear drives.

   公差设置不像齿轮传动装置那样严格。

7. Another advantage of belt drives is that they reduce vibration and shock transmission.

   皮带传动的另一个优点是可以减少振动及其冲击传递。

8. Furthermore, belt drives are relatively quiet.

   此外,皮带传动相对安静。

9. Some leather belts are in use at this time as well as flat steel, rubber, plastic, and fabric belts.

   除了平钢带、橡皮带、塑料带和布带外,一些皮革做的皮带也在使用。

10. Light, thin flat belts are practical on high speed machinery.

    轻质薄型平皮带适用于高速机械。

11. V belts are probably the most common means of transmitting power between electric motor and driven machinery.

    三角皮带大概是在电动机与从动机械之间传递动力的最普通的手段。

12. Conventional V belts are made of rubber.

    普通的三角皮带是用橡皮制造的。

13. Most often, both driver and driven pulleys lie in the same vertical plane.

    通常主动轮与从动轮位于同一个垂直面上。

14. Smooth flat belts and V belts depend on friction on the pulleys and some slippage is inherent in their operation.

    平皮带和三角皮带依靠皮带轮上的摩擦进行传动,所以在传动中总是有一些滑动。

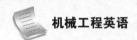机械工程英语

15. Most heavy duty pulleys are made of cast iron or formed steel.

    大多数的重型滑轮是用铸铁和型钢制造的。

16. Heat treatment is the operation of heating and cooling a metal in its solid state to change its physical properties.

    热处理是将金属在固态下加热和冷却以改变其物理性能的操作。

17. With the proper heat treatment internal stresses may be removed, grain size reduced, toughness increased, or a hard surface produced on a ductile interior.

    通过适当的热处理,可以清除内应力,细化晶粒,增强韧性,或在柔软材料上形成坚硬的表面。

18. The use of metals has always been a key factor in the development of the social systems of man.

    在人类社会的发展中,金属的应用起着关键性的作用。

19. At higher temperatures, the material is all austenitic, but as it cools it enters the region of ferrite and austenite stability.

    在较高温度时,这种材料全部都是奥氏体的,但当它冷却时,就会进入到铁素体和奥氏体稳定区。

20. Hyperthyroid steels are steels that contain greater than the eutectic amount of carbon.

    过共析钢是含碳量大于共析量的钢。

21. It should be remembered that the transitions that have been described by the phase diagrams are for equilibrium conditions, which can be approximated by slow cooling.

    要记住,由状态图描述的这种转化只适合于通过缓慢冷却的近似平衡条件。

22. Hardening is the process of heating a piece of steel to a temperature within or above its critical range and then cooling it rapidly.

    淬火就是把钢件加热到或超过它的临界温度范围,然后使其快速冷却的过程。

23. If the carbon content of the steel is known, the proper temperature to which the steel should be heated may be obtained by reference to the iron-iron carbide phase diagram.

    如果钢的含碳量已知,钢件合适的加热温度可参考铁碳合金状态图得到。

24. The heavier the section, the longer must be the heating time to achieve uniform results.

    截面越厚,加热的时间就要越长,才能达到均匀的结果。

25. Steel that has been hardened by rapid quenching is brittle and not suitable for most uses.

    快速淬火硬化的钢是硬且易碎的,不适合大多数场合使用。

26. This is usually accomplished by heating the steel too slightly above the critical temperature.

    这通常是通过将钢加热到比临界温度稍高一点来实现的。

27. Casting is a manufacturing process in which molten metal is poured or injected and allowed to solidify in a suitably shaped mold cavity.

    铸造是一种将熔化的金属倒入或注入合适的铸模腔并在其中固化的制造工艺。

28. Castings are parts that are made close to their final dimensions by a casting process.
通过铸造加工,铸件可以做得很接近它们的最终尺寸。

29. The molten material is poured into the pouring cup, which is a part of the gating system that supplies the molten material to the mold cavity.
熔化的金属从浇注杯注入型腔,浇注杯是浇注系统向型腔提供熔化金属的一部分。

30. Additionally there are extensions to the gating system called vents that provide the path for the built-up gases and the displaced air to vent to the atmosphere.
除此之外,还有称为排放口的浇注系统延长段,它可为合成气体和排出的空气提供通向大气的通道。

31. If two nodes transmit at the same time and interfere with each other's transmission, packets are corrupted.
如果两个节点同时传输,并且相互干扰,数据包就会被损坏。

32. The S-MAC protocol essentially trades used energy for throughput and latency.
S-MAC 协议基本上是用消耗的能量来换取吞吐量和延迟。

33. The firing of a periodic frame timer is very critical in the whole process of machinery.
在整个机械过程中,周期性帧定时器的触发非常重要。

34. When a node comes to life, it starts by waiting and listening.
当一个节点启动时,它从等待和侦听开始。

35. In the T-MAC protocol, every node transmits its queued messages in a burst at the start of the frame.
在 T-MAC 协议中,每个节点在帧开始时以突发的方式发送排队的消息。

36. Preliminary simulation experiments revealed a problem with the T-MAC protocol when traffic through the network is mostly unidirectional, like in a nodes-to-sink communication pattern.
初步的仿真实验表明,当通过网络的流量大多是单向(如节点到接收器的通信模式)时,T-MAC 协议存在一个问题。

37. We call the observed effect the early sleeping problem, since a node goes to sleep when a neighbor still has messages for it.
我们将观察到的效果称为早期睡眠问题,因为当一个节点进入睡眠状态时,相邻节点仍然有它的消息。

38. As the FRTS packet would otherwise disturb the data packet that follows the CTS, the data packet must be postponed for the duration of the FRTS packet.
由于帧中继流量整形包会干扰跟随 CTS 的数据包,数据包必须在帧中继流量整形包的持续时间内被延迟。

39. T-MAC uses a threshold: a node may only use this scheme when it has lost contention at least twice.
T-MAC 使用了一个阈值:一个节点只有在至少两次失去争用时才可以使用这个方案。

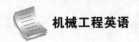

40. When a node is not transmitting, its radio is set to receive.

当节点不发送时,它的无线电被设置为接收。

41. So on this line, the S-MAC protocol is tuned to provide at least 90% throughput while using as little energy as possible for each individual load.

在这一行中,S-MAC 协议被调优为至少提供 90% 的吞吐量,同时对每个负载使用尽可能少的能源。

42. As with homogeneous local uni-cast we see that the maximum throughput of T-MAC is less than that of S-MAC.

与同质本地单播一样,我们发现 T-MAC 的最大吞吐量小于 S-MAC。

43. The maximum frequency that T-MAC can handle is lower than that of S-MAC, like we have seen in the nodes-to-sink communication pattern.

T-MAC 能够处理的最大频率低于 S-MAC,就像我们在节点到接收器的通信模式中看到的那样。

44. For the T-MAC protocol, we used overhearing avoidance, but no FRTS and no full-buffer priority.

对于 T-MAC 协议,我们使用了免监听,但是没有帧中继流量整形和全缓冲区优先级。

45. We have also not implemented the possibility to keep multiple schedules yet.

我们还没有实现保持多个时间表的可能性。

46. There is a transmission in most frames, but the radio is still in sleep mode most of the time.

大多数帧都有传输,但收音机大部分时间仍处于休眠模式。

47. We have only experimented with a static, non-mobile network.

我们只试验了静态的、非移动的网络。

48. Multi-hop messages travel at least two hops per frame because of T-MAC's in-band signaling: the CTS and ACK messages keep direct neighbors awake.

由于 T-MAC 的带内信令,多跳信息每帧至少传输两跳:CTS 和 ACK 信息使直接相邻节点保持清醒。

49. We start with a simple benchmark, where network load is homogeneous.

我们从一个简单的基准开始,网络负载是同质的。

50. To solve the problem of idle listening in a wireless sensor network, we have proposed the T-MAC protocol.

为了解决无线传感器网络中存在的空闲监听问题,我们提出了 T-MAC 协议。

Task Two　Sample Dialogue

Directions: *In this dialogue, you are going to read several times the following sample dialogue about the relevant topics. Please pay special attention to five Cs (culture, context, coherence, cohesion and critique) in the dialogue and get ready for a smooth communication in the coming* *task.*

The Car Broke Down

(*In this lesson, we will learn phrases you may use in talking about your car and related mechanic problems.*)

Burt: Hey Kevin! Where were you in the morning?

Kevin: Don't even talk about my morning.

Burt: Why? What happened? Is everything alright?

Kevin: My car broke down today and I had to miss such an important meeting.

Burt: Whoa! What happened to your car?

Kevin: The mechanic said that there was a problem with the radiator.

Burt: So, what about now? Did you get it repaired?

Kevin: Not yet, but the mechanic is taking care of it. I'll get it back by Tuesday.

Burt: So, don't worry. It's not a major problem. But how will you commute till Tuesday?

Kevin: Haven't thought about it yet.

Burt: Why don't we share a taxi until you get your car back?

Kevin: Thanks. I think that's a good idea.

Problem With the Car Battery

Kevin: Where are you, Mary?

Mary: I'm home. My car just won't start.

Kevin: I used to have the same problem with my old car. Did you check the battery?

Mary: Yeah. It was dead in the morning. I called the mechanic to give my car a jump.

Kevin: Well, I guess if you need to jump-start your car, you probably need to buy a new battery.

Mary: What are you saying? I bought this battery just a couple of months ago. I don't think I should have a problem with it every second day.

Kevin: I guess the dealer sold you an old one. Did he give you any warranty card with it?

Mary: Yes. I suppose.

Kevin: Then call the dealer and ask him to replace it. You are still eligible to claim the warranty.

Mary: All right. Thanks for your help. I'll call the dealer right now.

Task Three Simulation and Reproduction

Directions: *The class will be divided into three major groups, each of which will be assigned a topic. In each group, some students may be the teacher, while others may be students. In the process of discussion, please observe the principles of cooperation, politeness and choice of words. One of the groups will be chosen to demonstrate the discussion to the class.*

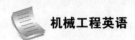

1) mechanical design in our daily life

2) a funny story related to mechanical design in my childhood

3) the importance of learning mechanical design

Task Four Discussion and Debate

Directions: *The class will be divided into two groups. Please choose your stand in regard to the following controversy and support your opinions with scientific evidences. Please refer to the specialized terms and classical sentences in the previous parts of this unit.*

At one time, engineers considered producing drawings as a job. Then along came CAD. The idea was to make the drawing faster and increase the number of projects completed over any given time period. Someone got the idea that all those draftsmen could be replaced by the engineer doing their own drawings. Do you agree?

V. After-class Exercises

1. *Match the English words in Column A with the Chinese meaning in Column B.*

A	B
1) shaft	A) 主轴
2) shear	B) 车床
3) stress	C) 剪切
4) mill	D) 载荷,负载
5) spindle	E) 轴
6) lathe	F) 氧化
7) load	G) 刚度
8) oxidation	H) 脱碳
9) rigidity	I) 应力
10) deodorization	J) 铣削

2. *Fill in the following blanks with the words or phrases in the word bank. Change the forms if it's necessary.*

pin	dampers	excitation	dimensions	cast
iterate	transformation	enveloping	tentatively	pattern

1) We will have to _____ to better and better solutions as we generate more information.

2) If a product configuration is _____ specified and then examined to determine whether the performance requirements are met.

3) Manufacturing can be defined as the _____ of raw materials into useful products through the use of the easiest and least-expensive methods.

4) As a result, the system will vibrate at the frequency of the _____ force regardless of the initial conditions or natural frequency of the system.

5) Before two components are assembled together, the relationship between the _____ of the mating surfaces must be specified.

6) The main practical advantage of lower pairs is their better ability to trap lubricant between their _____ surfaces.

7) Though frame material and design should handle damping, _____ are sometimes built into frame sections to handle specific problems.

8) Most frames are made from _____ iron, weld steel, composition, or concrete.

9) Although cast iron is a fairly cheap material, each casting requires a _____.

10) The term is used to describe joints with surface contact, as with a _____ surrounded by a hole.

3. *Translate the following sentences into English.*

1) 在物体上所画的基准线只会改变它的位置,不会改变它的角度方向。

_____

_____

2) 工程上的许多问题本质上是非线性的,也就是说恢复力并不与位移成正比,而且阻尼力也不与速度的一次方成正比。

_____

_____

3) 运动学仅研究机构的运动,而不考虑作用在机构上的力。

_____

_____

4) 当一个装置意外失效后,通常需要进行研究工作,找出失效的原因并确定可能的改正措施。

_____

_____

5) 一个或多个诸如齿轮、链轮、皮带轮和凸轮等类的构件通常借助于销、键、花键、卡环或其他装置连接到轴上。

_____

_____

4. *Please write an essay of about 120 words on the topic: **Application of mechanic design in our life**. Some specific examples will be highly appreciated and you have to watch out the spelling of some specialized terms you have learnt in this unit.*

_____

_____

_____

_____

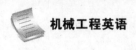

_____

_____

_____

_____

_____

_____

VI. Additional Reading

Engineering Design is the Culmination of Engineering Educational Process

Engineering design is a systematic process by which solution to the needs of humankind are obtained. The process is applied to problems (needs) of varying complexity. For example, mechanical engineers will use the design process to find an effective, efficient method to convert reciprocating motion to circular motion for the drive train in an internal combustion engine; electrical engineers will use the process to design electrical generating systems using falling water as the power source; and materials engineers use the process to design ablative materials which enable astronauts to safely reenter the earth's atmosphere.

The vast majority of complex problems in today's high technology society depend for solution not on a single engineering discipline, but on teams of engineers, scientists, environmentalists, economists, sociologists, and legal personnel. Solutions are not only dependent upon regulations andpolitical influence. As engineers, we have the technical expertise to develop products and systems, but at the same time we must be increasingly aware of the impact of our actions on society and the environment in general and work conscientiously toward the best solution in view of all relevant factors.

Design is the culmination of the engineering educational process; it is the salient feature that distinguishes engineering from other professions. A formal definition of engineering design is found in the curriculum guidelines of the Accreditation Board for Engineering and Technology (ABET). ABET accredits curricula in engineering schools and derives its membership from the various engineering professional societies. Each accredited curriculum has a well-defined design component which falls within the ABET statement on design.

Engineering design is the process of devising a system, component, or process to meet desired needs. It is decision making process (often iterative), in which the basic sciences, mathematics, and engineering sciences are applied to convert resources optimally to meet a sated objective, criteria, synthesis, analysis, construction , testing, and evaluation. The engineering design component of a curriculum must include most of the following features : development of student creativity, use of open-ended problem statement and use of modern design theory and methodology, formulation of design problem statement and specification, consideration of alternative solutions, feasibility considerations, production process, concurrent engineering design, and detailed system descriptions. Further, it is essential to include a variety of realistic constraints such as economic factors, safety, reliability, aesthetics, ethics, and social impact.

Drafting or technical drawing is the means by which mechanical engineers design products and create instructions for manufacturing parts. A technical drawing can be a computer model or hand-drawn schematic showing all the dimensions necessary to manufacture a part, as well as assembly notes, a list of required materials, and other pertinent information. A U. S. mechanical engineer or skilled worker who creates technical drawings may be referred to as a drafter or draftsman. Drafting has historically been a two-dimensional process, but computer-aided design (CAD) programs now allow the designer to create in three dimensions. Instructions for manufacturing a part must be fed to the necessary machinery, either manually, through programmed instructions, or through the use of a computer-aided manufacturing (CAM) or combined CAD/CAM program. Optionally, an engineer may also manually manufacture a part using the technical drawings, but this is becoming an increasing rarity, with the advent of computer numerically controlled (CNC) manufacturing. Engineers primarily manually manufacture parts in the areas of applied spray coatings, finishes, and other processes that cannot economically or practically be done by a machine. Working drawings are the complete set of standardized drawings specifying the manufacture and assembly of a product based on its design. The complexity of the design determines the number and types of drawings. Working drawings may be on more than one sheet and may contain written instructions called specifications. Working drawings are blueprints used for manufacturing products. Therefore, the set of drawings must: ①completely describe the parts, both visually and dimensionally(空间地;尺寸地);②show the parts in assembly;③identify all the parts; and ④specify the standard parts. The graphics and text information must be sufficiently complete and accurate to manufacture and assemble the product without error.

Generally, a complete set of working drawings for an assembly includes:

(1) detail drawings of each nonstandard part

(2) an assembly or subassembly(组件;子装配件) drawing showing all the standard and nonstandard parts in a single drawing

(3) a bill of materials (BOM)

(4) a title block

A detail drawing is a dimensioned(空间的;尺寸的), multiview(多视角;多视点) drawing of a single part, describing the part's shape, size, material, and surface roughness, in sufficient detail for the part to be manufactured based on the drawing alone. Detail drawings are produced from design sketches or extracted from 3D computer models. They adhere strictly to ANSI standards and the standard for the specific company, for dimensioning, assigning part

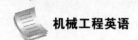 机械工程英语

numbers, notes, tolerances, etc.

In an assembly, standard parts such as threaded fasteners and bearings are not drawn as details, but are shown in the assembly view. Standard parts are not drawn as details because they are normally purchased, not manufactured, for assembly.

For large assemblies or assembled with large parts, details are drawn on multiple sheets, and a separate sheet is used for the assembly view. If the assembly is simple or the parts are small, detail drawings for each part of an assembly can be placed on a single sheet.

Multiple details on a sheet are usually drawn at the same scale. If different scales are used, they are clearly marked under each detail. Also, when more than one detail is placed on a sheet, the spacing between details is carefully planned, including leaving sufficient room for dimensions and notes.

1. *Read the passage quickly by using the skills of skimming and scanning, and choose the best answer to the following questions.*

    1) What's the main meaning of the passage? _____

       A. Engineering design is a systematic process.

       B. materials engineers' main work

       C. mechanical engineers' good ideas

       D. electrical engineers' main duty

    2) In the author's opinion, the design process is very complex and many factors should be taken into account. From the passage, who should NOT join into the design team? _____

       A. environmentalists   B. sociologists

       C. engineers   D. government officials

    3) From the passage, we can know the meaning of the word "ABET" (paragraph 3) is _____.

       A. a set of accredits standard

       B. a kind of design method

       C. a department of US government which is responsible for engineering design

       D. a set of law

    4) It can be concluded from the passage, which character of the following is NOT included in the curriculum? _____

       A. creativity   B. new material

       C. ethics   D. economic factors

    5) The title of the passage is _____.

       A. the roles of engineers in manufacturing

       B. the importance of mechanical design

       C. engineering design

D. the process of machine design

6）Based on the passage, which of the follow is wrong? _____

    A. Standard parts needn't drawn as details.

    B. For simple parts, sometimes they needn't to draw the detail working drawing.

    C. For large assemblies' details may be drawn on multiple sheets.

    D. If different scales used in a single sheet, they should be clearly marked under each detail.

7）Working drawings for an assembly MAY NOT include _____.

    A. BOM

    B. title block

    C. an assembly or subassembly drawing

    D. detail drawings of all parts.

8）For detail drawing, which one of the following statement is not true? _____

    A. should not have different scales

    B. Detail drawing can be produced from design sketches or extracted from 3D computer models.

    C. Detail drawing is a dimensioned, multivitamin drawing of a single part.

    D. should strictly adhere to the ANSI standards

9）From the passage we can infer ANSI is _____.

    A. a set of law   B. a set of notification

    C. an organization   D. IEEE government

10）The topic of the passage is _____.

    A. how to design working drawings

    B. working drawing's characters and the key influence of how to draw a working drawing

    C. the considered things in the design process

    D. the characters of working drawing

2. *In this part, the students are required to make an oral presentation on either of the following topics.*

1）the features of engineering design

2）the types and function of drawing

习题答案

Unit Two   Numerical Control

I. Pre-class Activity

Directions: *Please read the general introduction about **Richard Roberts** and tell something more about the great scientist to your classmates.*

Richard Roberts (1789—1864) was a British pattern maker and engineer whose development of high-precision machine tools contributed to the birth of production engineering and mass production. Roberts was born at Llanymynech, Powys, on the border between England and Wales.

Roberts built a range of machine tool, some to his own design, the first being a gear-cutting machine. For accurately checking the dimensions of the gears he adapted the sector, which he developed for sale to other engineers. Roberts adopted rotary cutters, which he had seen used at Maudslays. This is one of the earliest records of a milling cutter used in engineering. In 1816 he made the first reliable wet gas meter. In 1817 he made a lathe able to turn work 6 ft. long (1.8m). This had a back gear to give an increased range of speeds, and a sliding saddle to move the tool along the work. The saddle was driven by a screw through gearing which could be disengaged when the end of the cut was reached. Also in 1817 he built a planning machine to allow the machining of flat surfaces. According to biographer Richard Leslie Hills, his main contribution was the introduction of improved machine tools without which high standards of accuracy could not be achieved. This laid the foundation of production engineering as we know it today, leading to the interchangeability of standard parts and so mass production.

II. Specialized Terms

Directions：*Please memorize the following specialized terms before the class so that you will be able to better cope with the coming tasks.*

advance vt. (使)前进

analysis n. 分析；分解；验定

application n. 适用,应用

axes n. 轴

broadest adj. 宽广的

bur n. (多)刺

cam n. 凸轮

casting n. 铸造,铸件

clamp v. 夹紧,夹住

composite n. 复合材料；成物；菊科 vt. 使合成；使混合

continuum mechanics 连续介质力学

cycle n. 循环,周期；自行车；套；一段时间 vt. 使循环

discipline n. 学科；纪律；训练；惩罚 vt. 训练,训导；惩戒

distinction n. 差别,区别,不同

diversity n. 多样化

document n.文件,公文；[计]文档；证件

documentation n. 文件, 证明文件, 史实, 文件编制

doping n.[工程]上涂料

download vt.[计]下载

draft n. 汇票；草稿；选派

draw n. 绘画；制图

drive n. 驱动器；驾车；[心理]内驱力,推进力；快车道

driven adj. 被动的；收到驱策的；发奋图强的

dynamic adj. 动态的；动力的；动力学的；有活力的

dynamics n. 动力学,力学

engine n. 引擎,发动机；机车,火车头；工具

evaluate vt. 评价；估价；求……的值

file n. 锉刀

fluid mechanics[流]流体力学,液体力学

frame n.[电影]画面 vt. 设计；建造；陷害；使……适合

hoist n. 起重机,升降机

hook up 连接

hybrid system 混合系统

increment n. 增长；增量

inspecting and make-up machinery 检测及包装机械

intake n. 摄取量；通风口

integrate v. 整合；积分；集成化

intense adj. 强烈的；紧张的；非常的；热情的

jacquard n. 提花机

kinematics n. 运动学

knitting & hosiery machines 针织机及织袜机

knitting machines, over 165mm diameter 直径16毫米以上的圆形针织机

linear adj. 直线的

looms for elastic fabrics 弹性织物织机

machinist n. 机械师,机工

mechanics of material 材料力学

mechatronics n. 机电一体化；机械电子学

mercerizing machines for fabrics 织物丝光机

mercerizing machines for yarns 纱线丝光机

method n. 方法；条理；类函数

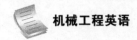

millimeter n. 毫米

milling n. 轧齿边

multiple adj. 多重的;多个的 n. 集成化;综合化

nanotechnology n. 纳米技术

non-woven fabric lines (dry process) 无纺织物生产线(干法)

non-woven fabric lines (wet process) 无纺织物生产线(湿法)

accessories and materials for knitting and hosiery machines 针织和袜机附件及材料

accessories for textile machines 纺织机附件

auxiliary machinery for spinning mills 纺纱厂辅助机械

auxiliary machines for finishing 整理用辅助机械

cutting machines 裁剪机

doubling and twisting machines 并捻联合机

machines for finishing 整理机

machines for spinning man-made fibres 化纤纺丝机

parts for spinning machines 纺纱机部件

parts for weaving machines 织机部件

preparatory machines for knitting 针织用准备机

overlap n. 重叠;重复 vi. 部分重叠;部分的同时发生

padding mangles 轧染机

patterning devices 提花设备

phase n. 时期

piston n. 活塞

porcelain accessories for textile machines 纺织机用陶瓷附件

positioning n. 定位

powerequipment and controlling equipment 电力设备和控制设备

preparatory machinery for cotton spinning systems 棉纺前纺机械

principle n. 原理,原则;主义

printing machinery 印花机械

Rapier looms 剑杆织机

Raschel looms 拉舍尔经编机

rectangular adj. 矩形的;成直角的

retaliation n. 报复;反击;回敬

revolution n. 革命;旋转;运行;循环

robotic adj. 机器人的,像机器人的;自动的 n. 机器人学

rotary adj. 旋转的

rotary screen printing machines 圆网印花机

rotor spinning frames 气流纺纱机

sewing machines 缝纫机

software n. 软件

spinning machinery 纺纱机

statics n.静力学

tensionless dryers 无张力干燥机

tentering and stentering machines 拉幅机

thermo-fixing machines 热定形机

top printing machines 套色印花机

traced v. 追溯(trace 的过去分词);跟踪;摹写 adj. 示踪的;摹写的

typical adj. 典型的;特有的;象征性的

vehicle n.[车辆]车辆;工具;交通工具;运载工具;传播媒介;媒介物 vt. 与……重叠;与……同时发生

warp stop motions 断经自停装置

warping machines 整经机

weapon n. 武器,兵器

wrench n. 扳手

III. Watching and Listening

视频链接及文本

Task One   A CNC Machinist

New Words

machinist n. 机械师, 机工

casting n. 铸造, 铸件

hook up 连接

file n. 锉刀

bur n. (多) 刺

hoist n. 起重机, 升降机

clamp v. 夹紧, 夹住

multiple adj. 多重的; 多个的

wrench n. 扳手

advance vt. (使) 前进

Exercises

1. *Watch the video for the first time and choose the best answer to the following questions.*

1) What are the main jobs of Walter Wood? _____

   A. taking a casting model

   B. making casting into a finished product

   C. running the machine

   D. hooking up to a computer

2) What is the typical day of Walter Wood? _____

   A. making sure that his machine is safe

   B. making his area clean

   C. making sure everything he runs is correct

   D. the above three options are correct

3) What's the first thing Walter Wood often does? _____

   A. to check the rough casting out

   B. to take a file to file the burs off

   C. to lift the casting up with the hoist

   D. to clamp the rough casting into the jaws

4) What are Walter Wood's most valuable skills? _____

   A. reading the tools to get the part on size

   B. reading the tools and geting the part on size

   C. knowing if his part is on size

   D. running multiple pieces, multiple castings

5) Where did Walter Wood learned mechanical skills? _____

   A. in the factory   B. in high school

   C. in CNC   D. in a metals class

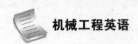
2. *Watch the video again and decide whether the following statements are true or false.*

1）This program will concentrate mainly on the basic CNC principles. （　）

2）My main responsibility is to take a rough casting. （　）

3）The first thing I do is to check the rough casting out. （　）

4）My most valuable skill, I would say, would have to be reading the books. （　）

5）Every day is something different. （　）

3. *Watch the video for the third time and fill in the following blanks.*

My main responsibility is to take a _____ casting, and turn it into a _____ product. A CNC machine is _____ a machine that is hooked up to a _____ that runs off of a _____. And that program is going to tell that machine what type of _____, what type of _____, to cut on that part. A _____ day for me is: making sure that my machine is _____, my area is clean, and making sure that whatever part I'm running, I run it right, and I break it on size, and make it a _____ part.

4. *Share your opinions with your partners on the following topics for discussion.*

1）Do you know the responsibilities of a CNC machinist?

2）Can you turn a rough casting into a finished product?

Task Two　Computer Numerical Control

New Words

diversity n. 多样化	rotary adj. 旋转的
application n. 适用,应用	positioning n.定位
milling n. 轧齿边	rectangular adj.矩形的；成直角的
linear adj. 直线的	increment n. 增长；增量
axes n. 轴	millimeter n. 毫米

视频链接及文本

Exercises

1. *Watch the video for the first time and choose the best answer to the following questions.*

1）Why does this program will concentrate primarily on the basic computer numerical control principles of vertical milling machining center? _____

A. because of the application of machine tool

B. because of the types of machine tool

C. because of the diversity of machine tool

D. because of the diversity of machine tools, types and applications

2）Which is called the y-axis? _____

A. the table's motion side to side

B. the table's motion in and out

C. the head movement up and down the column

D. the fourth rotary axis

3) Which represents the right angle perpendicular to both the x and y axes? _____

   A. the y axis   B. the a axis

   C. the z axis   D. the x axis

4) For machines using metric, the smallest increment is _____ .

   A. one ten-thousandth of an inch

   B. one thousandth of a millimeter

   C. one thousandth of an inch

   D. usually ten thousandth of a millimeter

5) Which category of the writing belongs to? _____

   A. political   B. economic

   C. cultural   D. science

2. *Watch the video again and decide whether the following statements are true or false.*

   1) Due to the variety of machine tools, types and applications, this program will mainly focus on the computer numerical control principle of vertical milling processing center. (   )

   2) Vertical milling centers usually have four straight axes. (   )

   3) If a rotation table is added to the platform, there is a fourth rotation axis, called a axis. (   )

   4) All baseline increments are specified in linear measurements. For most machines that use the English system, the smallest increase is one tenth of an inch. (   )

   5) A circle coordinate system allows mathematical drawing of points in space, called coordinates between two or more machine axes. (   )

3. *Watch the video for the third time and fill in the following blanks.*

   If a rotary table is added on the machine table, then there is a _____ rotary axis which is called the a axis. The method of accurate _____ along each _____ for all machine types is achieved using the _____ coordinate system. On the _____ milling machining center, the _____ base line of the rectangular coordinate system represents the x axis, and the vertical base line represents the y axis, the z axis is at a right _____ perpendicular to both the x and y axes. Increments for all baselines are specified in _____ measurement. For most machines using the English system, the smallest _____ is one ten-thousandth of an _____ .

4. *Share your opinions with your partners on the following topics for discussion.*

   1) How do you identify the axis of motion on the vertical milling machining center? Please summarize the ways of positioning along each axis.

   2) Can you use a few lines to list what's your understanding about CNC? Please use an example to clarify your thoughts.

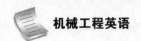

IV. Talking

Task One   Classical Sentences

Directions：*In this section, some popular sentences are supplied for you to read and to memorize. Then, you are required to simulate and produce your own sentences with reference to the structure.*

General Sentences

1. What's your nationality? Are you Chinese?
   你是哪国人？是中国人吗？

2. What part of the world do you come from?
   你来自哪里？

3. I was born in Spain, but I'm a citizen of France.
   我出生在西班牙，但我是法国公民。

4. Do you know what the population of Japan is?
   你知道日本有多少人口吗？

5. What's the area of the Congo in square miles?
   刚果的面积是多少平方英里？

6. Who is the governor of this state?
   谁是这个州的州长？

7. According to the latest census, our population has increased.
   根据最新的人口普查，我们的人口增加了。

8. Politically, the country is divided into fifty states.
   该国从行政上被划分为 50 个州。

9. The industrial area is centered largely in the north.
   工业区大部分集中在北方。

10. The country is rich in natural resources. It has large quantity of mineral deposits.
    该国自然资源丰富，有大量的矿藏。

11. This nation is noted for its economic stability.
    该国以经济稳定闻名。

12. The U. S. is by far the biggest industrial country in the world.
    到目前为止，美国是世界上最大的工业国。

13. My home is in the capital. It's a cosmopolitan city.
    我的家在首都，它是一个国际大都会。

14. Geographically, this country is located in the southern hemisphere.
    从地理位置上讲，这个国家位于南半球。

15. Britain is an island country surrounded by the sea.

英国是一个被海环绕的岛国。

16. It's a beautiful country with many large lakes.

这是一个有着若干大湖的美丽国度。

17. Thames River is the second longest and the most important river in Britain.

泰晤士河是英国第二大河,也是英国最重要的河。

18. This part of the country is very mountainous.

这个国家的这部分被众多山脉覆盖。

19. Mount Tai is situated in the western Shandong province.

泰山位于山东省西部。

20 Along the northern coast there are many high cliffs.

北海岸多危壁断崖。

21. There are forests here, and lumbering is important.

此地多森林,故以伐木业为重。

22. In Brazil, many ancient forests are very well preserved.

在巴西,很多古老的森林保存十分完好。

23. The scenery is beautiful in the small islands in the Pacific Oceans.

太平洋上一些小岛的景色十分优美。

24. This mountain range has many high peaks and deep canyons.

此山高峰深谷众多。

25. What kind of climate do you have? Is it mild?

你们那里气候怎么样? 温和吗?

26. Britain has a maritime climate—winters are not too cold and summers are not too hot.

英国属于海洋性气候,冬季不过于寒冷,夏季不过于炎热。

27. The land in this region is rather dry and parched.

这片土地十分干燥。

28. How far is it from the shore of the Atlantic to the mountains?

从大西洋海岸到山区有多远?

29. Lumbering is very important in some underdeveloped countries.

在一些不发达国家,伐木业十分重要。

30. What's the longest river in the United States?

美国最长的河是什么河?

31. Are most of the lakes located in the north central region?

大部分湖泊是不是在北部的中心地区?

32. As you travel westward, does the land get higher?

你去西部旅行时,是不是地势越来越高?

33. The weather is warm and sunny here. Do you get much rain?

这里的天气温暖而晴朗。多雨吗?

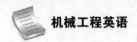

34. Because of the warm and sunny weather, oranges grow very well here.
因为这里气候温暖,光照充足,橘子长势很好。

35. In this flat country, people grow wheat and corn and raise cattle.
这个国家地势平坦,人们种植小麦、玉米,饲养牲畜。

36. The ground around here is stony and not very good for farming.
这周围的土地多石,不适合耕种。

37. Is the coastal plain good for farming?
这种海边的平原有利于发展农业吗?

38. Is the plain along the river good for farming?
河畔的平原易于发展农业吗?

39. What are the principal farm products in this region?
这个地区的主要农产品是什么?

40. Milk, butter, and cheese are shipped here from the dairy farms.
牛奶、黄油、奶酪都从奶制品农场运到这里。

41. They had to cut down a lot of trees to make room for farms.
他们不得不砍伐一些树木,从而为农场提供足够的空间。

42. At this time of the year farmers plow their fields.
一年中的这个时候农民们会耕种自己的土地。

43. On many farms you'll find cows and chickens.
在许多农场你都能发现奶牛和鸡。

44. If you have cows, you have to get up early to do the milking.
如果你有奶牛,你得早起挤牛奶。

45. Tractors have revolutionized farming.
拖拉机使农业发生了革命性的变化。

46. In the United States, there are many factories for making cloth.
美国有很多制布厂。

47. Factories employ both male and female workers.
工厂既雇用男工,也雇用女工。

48. If you work in a factory, you usually have to punch a clock.
如果你在工厂工作的话,你就得打卡。

49. Is meat packing a big industry in your country?
肉类包装在你们国家是不是一个大的产业?

50. Is it true that manufacturing of automobiles is a major industry?
汽车制造业是主要产业,是吗?

Specialized Sentences

1. A lathe is a machine tool used primarily for producing surfaces of revolution and flat edges.
车床主要用于制作旋转表面和平整边缘的机床。

2. The mold is made by making a pattern using wax or some other material that can be melted away.

制作铸型的型模采用石蜡或其他一些能融化的材料做成。

3. Attention must be paid to mold permeability when using pressure, to allow the air to escape as the pour is done.

当使用压力时必须注意模具的渗透性,以便在浇铸的同时让空气逸出。

4. The low carbon content permits so little formation of hard martensite that the process is relatively ineffective.

碳含量太低,很难形成硬的马氏体结构,从而使热处理相对不起作用。

5. Types of parts that can be made using these processes are pump housings, manifolds, and auto brake components.

使用这些工艺制造的零件类型有泵壳、集合管和自动制动部件。

6. Forging is the process by which metal is heated and is shaped by plastic deformation by suitably applying compressive force.

在锻造过程中先将金属加热,然后施加合适的压力使其塑性变形。

7. The feature common to all of them is that, like the hand forging hammer.

它们的共同特点,都像手工锻锤一样。

8. They utilize the energy of a falling weight to develop the pressure needed for shaping the metal.

它们利用落锤能量来产生金属成型所需的压力。

9. One of the basic operations of hammer forging is the elongation of a piece of metal by stretching with hammer blows, causing it to become thinner and longer.

锤锻的基本操作之一就是通过锤击使金属伸长,使其变细变长。

10. More important is closed-die forging, very widely used for mass production in industry.

更重要的是闭式模锻,在工业上广泛用于规模生产。

11. The metal is shaped by pressing between a pair of forging dies

金属在一对锻模之间压制成型。

12. Forgings yield parts that have high strength to weight ratio, thus are often used in the design of aircraft frame members.

锻造生产的零件具有较高的强度重量比,所以常被用在飞机结构零件的设计中。

13. The importance of machining processes can be emphasized by the fact that every product we use in our daily life has undergone this process either directly or indirectly.

机械加工过程的重要性,可通过日常生活使用的每件产品直接或间接经历这一过程的事实来强调。

14. Machining operations are utilized in view of the better surface finish that could be achieved by it compared to other manufacturing operations.

由于机械加工能获得比其他制造作业更好的表面光洁度,所以机械加工作业具有实用价值。

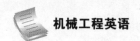

15. Grinding is a manufacturing process that involves the removal of metal by employing a rotating abrasive wheel.

磨削是通过采用旋转磨轮去除金属的制造工艺。

16. Generally, grinding is considered to be a finishing process that is usually used for obtaining high-dimensional accuracy and better surface finish.

一般而言,磨削被认为是一种通常用于获得高尺寸精度和较好表面光洁度的精加工作业。

17. It can be used to make parts that cannot be produced by normal manufacturing techniques.

它还能用于生产一般制造技术无法生产的零件。

18. Drilling operations can be carried out by using either hand drills or drilling machines.

钻削作业既能采用手钻,也能采用钻床来实现。

19. Milling is a machining process that is carried out by means of a multi-edge rotating tool known as a milling cutter.

铣削是一种利用被称为铣刀的多刃旋转刀具进行的加工过程。

20. Turbine blades that have complex shapes, or airplane parts that have to withstand high temperatures.

形状复杂的涡轮叶片或必须耐得住高温的飞机零件。

21. This rotation of both the wheels would result in slipping of one or both of them on the road surface.

这样转动将导致一只或两只后轮在路面上打滑。

22. Neither the misalignment resulting from the flexing of the vehicle structure over bumpy road surfaces can be avoided nor the precise alignment of shaft can be ensured without them.

没有万向节,不仅因颠簸路面引起车身结构弯曲造成的不同轴线不能避免,而且轴的精确同轴也不能保证。

23. Cores are placed in molds wherever it is necessary to preserve the space, it occupies in the mold as a void in the resulting castings.

型芯可放在铸型中,占据空间,以便在最后的铸件中形成空间。

24. Green sand generally consists of silica sand and additives coated by rubbing the sand grains together with clay uniformly welted with water.

型砂通常含有石英砂和添加剂,添加剂通常是将砂粒与黏土均匀地与水一起搅拌,使沙粒及添加剂表面包覆一层黏结薄膜。

25. As a result, the last portion of the casting to freeze will be deficient in metal, and in the absence of a supplemental metal-feed source, will result in some form of shrinkage.

结果,在铸件最后凝固的部分会缺少金属,而且在缺少补缩金属源时,将会产生一些缩孔。

26. The pouring process must be carefully controlled.

浇注过程必须小心控制。

27. Again, productivity and quality depend to a large degree on the skill of the hammer operator and the tooling.

同样地,生产率和质量很大程度上取决于锻锤操作者的技巧和所用的工具。

28. Closed-die forging relies less on operator skill and more on the design of the preform and forging dies.

封闭模锻造很少依赖操作者的技巧,而更多地取决于预成型模和锻造模的设计。

29. Tolerances vary with materials and design, but production runs calling for 0.002 to 0.005 in tolerance are regularly made.

公差随材料和设计而变,但生产上通常需要 0.002~0.005 英寸的公差。

30. Ring rolling offers a homogeneous circumferential grain flow, ease of fabrication and machining, and versatility of material size.

环状轧制可得到均质的周向晶粒流,易于制造和加工,可用于多种材料尺寸。

31. It also incorporates the "chattering vibration" phenomenon in the actuators, which must be avoided in many physical systems such as servo control systems, structure vibration control systems, and robotized systems.

它还包含了执行器中的"颤振现象",必须在许多物理系统中避免振动现象,如伺服控制系统、结构振动控制系统和机器人系统。

32. For that reason a modification of the classical SMC is introduced through the hyperbolic tangent function, with the purpose of reducing its characteristic abrupt commutation, as indicated in equation.

为此,采用双曲正切函数对经典 SMC 进行了修正,以减少其特征突变,如方程所示。

33. The simulation tools are the foundation for the design of robot systems, for the application of robots in complex environments and for the development of new control strategies and algorithms.

仿真工具为机器人系统的设计、机器人在复杂环境中的应用以及新的控制策略和算法的开发奠定了基础。

34. The design, simulation and comparison of the performance of controllers applied to a redundant robot with five degrees of freedom(DOF) are presented in this paper.

文章对五自由度冗余机器人的控制器性能进行了设计、仿真和比较。

35. Six controllers are prepared to test the robot's dynamic model: hyperbolic sine-cosine; computed torque; sliding hyperbolic mode; control with learning and adaptive.

准备六个控制器测试机器人的动力学模型:双曲正余弦;计算扭矩;滑动双曲模式;学习控制;自适应控制。

36. Process planning in special machinery: increasing reliability in volatile surroundings.

特殊机械的工艺规划:在多变的环境中提高可靠性。

37. Special machinery solutions gain more importance for German machinery constructors.

特殊机械解决方案越来越受到德国机械制造商的重视。

38. The main task in the commissioning process is to establish the functionality and the

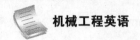

functional interaction of previously assembled components as well as their testing.

调试过程中的主要任务是建立以前组装的组件的功能和功能交互,以及它们的测试。

39. The commissioning process is challenged by complexity, time pressure and concurrency of errors.

复杂性、时间压力和误差的并发性对调试过程提出了挑战。

40. With the first alignment of different components the technical problems usually increase.

随着不同组件的首次对准,技术问题通常会增加。

41. These problems are cost-intensive and time-consuming and can cause delays.

问题是成本密集和耗时,可能导致延误。

42. Paired with the long lead times, high capital costs and the complexity in special machinery, the commissioning process represents a high saving potential for special machinery manufacturers.

由于特殊机械交货期长、资金成本高、复杂程度高,调试过程对特殊机械制造商来说具有很高的节约潜力。

43. While a large amount of studies focused on the commissioning process along the ramp-up of serial production, very little research is conducted on this process in small-scale production environments.

虽然大量的研究都集中在批量生产的调试过程上,但是很少有研究是在小型生产环境中进行的。

44. Furthermore, the commissioning process accounts for capital commitment costs as well as a shortage of available space on the shop floor.

此外,调试过程包括资本承诺成本以及车间可用空间的不足。

45. Development of a generic module for the commissioning process in special machinery.

开发专用机械调试过程中的通用模块。

46. When special machinery manufacturers analyze their processes in order to gain improvements regarding time, cost or quality aspects, they often need to establish changes.

当特殊机械制造商分析他们的过程以获得时间、成本或质量方面的改进时,他们经常需要建立变化机制。

47. Depending on the scope of the process changes, the impact is often hard to evaluate previously.

根据流程更改的范围,影响通常很难预先评估。

48. Given the small quantities, the long lead times and the cost intensity of customized machines, prototyping does not seem to be a suitable option in every case.

考虑到订制机器的数量少、交货期长和成本强度,原型设计似乎并不是所有情况下都合适的选择。

49. It is important to simulate the changing process when a difficult task emerges.

当出现困难任务时,模拟变化的过程很重要。

50. This step is crucial for the success of the simulation as an inadequate simulation model holds the risk of misleading conclusions.

这一步对模拟的成功是至关重要的,因为一个不适当的模拟模型有误导结论的风险。

Task Two   Sample Dialogue

Directions: *In this dialogue, you are going to read several times the following sample dialogue about the relevant topic. Please pay special attention to five Cs (culture, context, coherence, cohesion and critique) in the dialogue and get ready for a smooth communication in the coming task.*

(*Topical introduction : Tom and Mary have now just finished a class given by Professor Smith on the workings of the machine tools. Yet, they still have some questions, and so, they walk over to Prof. Smith.*)

Tom: (to Prof. Smith) Excuse me, sir. I'm Tom, one of your students in this class. Could I ask you some questions?

Smith: (to Tom): Why not? Come on, then.

Tom: Why do we name these machines numerical control machines?

Smith: A good question, boy. You know, uh-hm... NC, or numerical control, actually refers to the control of a machine tool or any other processing machines by using a series of mathematical information, or numerical data. It means the work of machine is controlled by a numerical control program.

Mary: Oh, Prof. Smith, the idea is nice enough. But what is the advantage of numerical control over the hand control? Isn't our hand more sensible or reliable than the programmed data?

Smith: In some ways, yes. A good example is some fine works, like a jade box, but the qualities of hand-made products may not be consistent or stable. What is more important, NC has proved to be much more advantageous in overall operation.

Mary: But what do you mean by overall operation?

Smith: Uh-huh, by it we mean the general industrial production practice. For example, NC should be adopted whenever there is similar raw material and work parts are produced in various sizes and complex shapes. Those production shops that may have frequent changeovers will surely benefit from NC programs.

Tom: Thank you, Prof. Smith, but I have another question. How can we use NC to get more satisfactory results in the real production?

Smith: It's better to use NC tools together with other technical advances, such as programmed optimization of cutting speeds and feeds, work positioning, tool selection, and chip disposal.

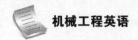

Tom & Mary：Thank you Prof. Smith. NC sounds really promising from what we say.

Task Three Simulation and Reproduction

Directions：*The class will be divided into three major groups, each of which will be assigned a topic. In each group, some students may be the teacher, while others may be students. In the process of discussion, please observe the principles of cooperation, politeness and choice of words. One of the groups will be chosen to demonstrate the discussion to the class.*

1) computer numerical control in our daily life
2) a funny story related to hand-made products in my childhood
3) the importance of learning computer numerical control

Task Four Discussion and Debate

Directions：*The class will be divided into two groups. Please choose your stand in regard to the following controversy and support your opinions with scientific evidences. Please refer to the specialized terms and classical sentences in the previous parts of this unit.*

Numerical control (NC) is the automation of machine tools that are operated by precisely programmed commands encoded on a storage medium, as opposed to controlled manually by hand wheels or levers, or mechanically automated by cams alone. Can you give more samples of NC today? Such as computer numerical control (CNC).

V. After-class Exercises

1. *Match the English words in Column A with the Chinese meaning in Column B.*

A	B
1) view	A) 工件
2) machining	B) 磨床
3) found	C) 加工
4) assembling	D) 铣刀
5) precision	E) 装配
6) work-piece	F) 视图
7) milling cutter	G) 精度
8) grinder	H) 铸造
9) stamping	I) 压模
10) axes	J) 轴

2. *Fill in the following blanks with the words or phrases in the word bank. Change the forms if it's necessary.*

| milling cutter | lathe | commands | hand drills | metal |
| finish | instructions | punched | accuracy | tooling |

1) Drilling operations can be carried out by using _____ .

2) Milling is a machining process that is carried out by means of a multi-edge rotating tool known as a _____ .

3) Generally, grinding is considered to be a finishing process that is usually used for obtaining _____ .

4) A _____ is a machine tool used primarily for producing surfaces of revolution and flat edges.

5) Numerical control (NC) is the automation of machine tools that are operated by precisely programmed _____ .

6) The numerical control machine is defined as the machine that is controlled by the set of _____ called as the program.

7) In the late 1940s Parsons adopted the method of using _____ cards containing coordinate position system to control a machine tool.

8) Machining operations are utilized in view of the better surface _____ .

9) Productivity and quality depend on the skill of the hammer operator and the _____ to a large degree.

10) The NC technology can be more prominently used for various _____ machining processes.

3. *Translate the following sentences into English.*

1) 万能加工中心同时装备了垂直主轴和水平主轴的机器。

_____

2) 数控是程序控制的自动化,在数字控制系统中,设备通过数字、字母和符号来编码。

_____

3) 线性差补是最基本的差补方法,用于连续路径的数控系统中。

_____

4) 直接数据控制定义为一个制造系统,一定数量的机床由一台计算机通过直接硬件连线实时控制。

_____

5) 计算机辅助编程是一种涉及特殊符号的编程语言,这种语言可以决定点的坐标、刀口以及工件的表面。

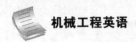

机械工程英语

4. *Please write an essay of about 120 words on the topic*: **Application of NC or CNC in our life.** *Some specific examples will be highly appreciated and you have to watch out the spelling of some specialized terms you have learnt in this unit.*

_____
_____
_____
_____
_____
_____
_____
_____
_____
_____
_____

VI. Additional Reading

Numerical control, popularly known as the NC is very commonly used in the machine tools. Numerical control is defined as the form of programmable automation, in which the process is controlled by the number, letters, and symbols. In case of the machine tools this programmable automation is used for the operation of the machines.

In other words, the numerical control machine is defined as the machine that is controlled by the set of instructions called as the program. In numerical control method the numbers form the basic program instructions for different types of jobs; hence the name numerical control is given to this type of programming. When the type of job changes, the program instructions of the job also change. It is easier to write the new instructions for each job, hence NC provides lots of flexibility in its use.

The NC technology can be applied to wide variety of operations like drafting, assembly, inspection, sheet metal working, etc. But it is more prominently used for various metal machining processes like turning, drilling, milling, shaping, etc. Due to NC all the machining operations can be performed at the fast rate resulting in bulk manufacturing becoming quite cheaper.

The invention of numerical control has been due to the pioneering works of John T. Parsons in the year 1940, when he tried to generate a curve automatically by milling cutters by providing coordinate motions. In the late 1940s Parsons conceived the method of using punched cards containing coordinate position system to control a machine tool. The machine directed to move in small increments and generate the desired finish. In the year, 1948, Parons demonstrated this concept to the US Air Force, who sponsored the series of project at laboratories of Massachusetts Institute of Technology (MIT). After lots of research MIT was able to demonstrate first NC prototype in the year 1952 and in the next year they were able to prove the potential applications of the NC.

Soon the machine tool manufacturers began their own efforts to introduce commercial NC units in the market. Meanwhile, the research continued as MIT, who were able to discover Automatically Programmed Tools, known as APT language that could be used for programming the NC machines. The main aim of APT language was to provide the means to the programmer by which they can communicate the machining instructions to the machine tools in easier manner using English-like statements. The APT language is still used widely in the manufacturing industry and a number of modern programming languages are based on the concepts of APT.

In the initial years of NC, punched tapes were for feeding the instructions to the machine tools via the control unit. The APT language also marked the arrival of the computer numerical controlled machines, popularly known as the CNC machines. Another language, PRONTO, was discovered by Parick Hanratty, who carried out various experiments at GE and released the language in the year 1958.

In CNC machines programs are fed in the computer, which was used to control the operations of the machines. Thus the control unit used that would read the punched cards in the NC machines was replaced by the microcomputer in the CNC machines. The CNC brought major revolution in the manufacturing industry. The next development has been the combination of computer aided manufacturing (CAM) and computer aided designing (CAD) called as CAD/CAM.

1. *Read the passage quickly by using the skills of skimming and scanning, and choose the best answer to the following questions.*

   1) What does the numerical control process include? _____

      A. number   B. letters

      C. symbols   D. all the above

   2) What is the programmable automation used for? _____

      A. set of instructions   B. operation of the machines

      C. as a tool   D. control of the machine

   3) What's the correct understanding of a machine tool? _____

      A. The machine that is controlled by the set of instructions called as the program.

      B. a kind of process

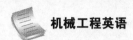

C. an instruction tool

D. a machine controlled by a set of instructions

4) When will the program instructions for the job change? _____

A. when the machine tool is damaged

B. when the machine is broken

C. when the job type changes

D. when the basic program instructions are different

5) Why it is easier to write a new instruction for a job? _____

A. because the technology is so advanced

B. because NC provides a lot of flexibility in use

C. because it's so simple

D. because designers are smart

6) In what aspect is the application of numerical control more prominent? _____

A. drafting

B. assembly

C. inspection and sheet metal working

D. various metal machining processes like turning, drilling, milling and shaping

7) When was the first CNC prototype shown? _____

A. 1940   B. 1948

C. 1952   D. 1925

8) What is called APT language? _____

A. the machining instructions to the machine tools

B. computer language

C. numerical control machine programming

D. CNC prototype

9) What's the main purpose of APT language? _____

A. for manufacturing application

B. for transferring instructions

C. to provide programmers with more convenient means of transmitting instruction to machine tools

D. to automatically discover programming

10) Why is the control unit for reading punch cards in NC machine tools replaced by microcomputers in NC machine tools? _____

A. because it is so outdated

B. because programs are used on the computer to control the operation of the machine

C. because the numerical control technology has brought the significant revolution to the manufacturing industry

D. because Parick Hanratty found another language

2. *In this part, students are required to make an oral presentation on either of the following topics.*

1) the features of CNC

2) the application of CNC machines in engineering

习题答案

Unit Three Advanced Manufacturing Technologies

I. Pre-class Activity

Directions: *Please read the general introduction about* **Martin Wiberg** *and tell something more about the great scientist to your classmates.*

Martin Wiberg (1826—1905) was born in Viby, Scania, Sweden, enrolled at Lund University in 1845 and became a Doctor of Philosophy in 1850. He is known as a computer pioneer for his invention of a machine that could print logarithmic tables (first interest tables appeared in 1860, logarithmic in 1875). The logarithmic tables were subsequently published in English, French and German in 1876. The device was investigated by the French academy of science which also wrote an extensive report on it in 1863. The device was inspired by the similar work done by Per Georg Scheutz (had the same capacity: 15-digit numbers and fourth-order differences) and has similarities with Charles Babbage's difference engine. (Scheutz machine was based on the difference engine)

Wiberg was an outstanding inventor, whose contributions ranged from mechanical letterboxes, heating devices for railroad carriage components, speed controls, match-manufacturing machines to self-propelled torpedoes, and automatic breech loading weapons, a cream separator, etc. His calculator became especially well-known through a set of logarithm tables, which included logarithms of the trigonometric functions, and appeared in 1875. It was published in Swedish, German, French and English editions, appearing in 1876, in time to be included among Sweden's contributions to the International "Centennial" Exposition held at Philadelphia that year, where it joined his "bull-dog apparatus" for deep-sea sounding and railway control device.

II. Specialized Terms

Directions: *Please memorize the following specialized terms before the class so that you will be able to better cope with the coming tasks.*

AC circuit 交流电路

aerospace n. 航空与航天

allocation n. 定位

anneal n. 退火

bending stress n. 弯曲应力

circuit n. 电路

composite n. 复合材料

composition n. 构图

cross-section n. 横断面

cutter n. 刀具

decarburization n. 脱碳

deformation n. 变形

deforming force n. 变形力

diffusion n. 扩散

dynamic n. 动力学

elasticity n. 弹性

error n. 误差

external grinding 外圆磨削

extruder n. 挤压机

fabrication n. 制造

fiber n. 光纤

fraising n. 绞孔

frequency n. 频率

friction n. 摩擦

functional adj. 功能的

gearbox n. 变速箱

geometrical n. 几何形状

geometry n. 几何形状

heat treatment 热处理

hitting v. 压缩

instrument n. 仪器

intensity n. 强度

internal grinding 内圆磨削

jig n. 机床夹具

kinematic n. 运动学

link n. 联结

logistics n. 后勤

machine tool 机床

magnetic circles 磁路

manufacturing n. 工厂

mentor n. 有经验的顾问

module n. 模块；组件

normalizing n. 正火

planarization v. 平面化

plane grinding 平面磨削

polymers n. 聚合物

processing power 处理能力

processor n. [计]处理器；处理程序；加工者

produce v. 生产；创造

product success rate 产品(研发/设计的)成功率

product test 产品检验,产品试验

program n. 程序；计划；大纲

progression n 连续

properties n. 特性

property n. [复数 properties]性质,性能；财产；所有权

prototyping n. 原型机制造

P-type semi-conductor P 型半导体

pulling v. 拉伸

purify vt. 净化；使纯净 vi. 净化；变纯净

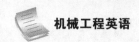

quantities n. 数量;工程量;音符的长度
（quantity 的复数）

radio access 无线接入

radio interface 无线电接口,无线接口

random access memory(abbr. RAM) 随机存取存储器

read-only memory 只读内存

reamer n. 绞刀

rectification n.改正,正;[化工]精馏;[电]整流;[数]求长

reference n. 参照,参考

refine vt. 精炼,提纯;改善

regulate vt. 控制,管理

relay n. 继电器

remarkable adj. 出色的,非凡的

remote access 远程访问;远程存取

remote terminal 远程终端

rentilation n.房租

research and development department 研发部门

resistance n. 阻力;电阻;抵抗;反抗;抵抗力

response v. 响应

restock vt. 重新进货;再储存 vi. 补充货源;补足

revolution n. 革命;旋转;运行;循环

revolutionize vt. 发动革命;彻底改革;宣传

革命

rotate v. 轮换;交替

rich-media 富媒体

rigidity n. 硬度

semiconductor n. <物>半导体

shear n. 剪切

shaft n. 轴

sinusoid adj. 正弦形的

solid adj. 固体的;实心的

static n. 静力学

stress n. 应力

thermoplastic adj. 热塑性的

thread processing 螺纹加工

transparent adj. 透明的

twist n. 扭转

wideband adj. 可用多种频率的,宽频带的

WiFi abbr. (wireless fidelity) 无线上网技术

wind power generation 风力发电

wireless adj. 无线的;无线电的 n. 无线电 vt. 用无线电报与……联系;用无线电报发送 vi. 打无线电报

wireless broadband access 无线宽带接入

wireless network 无线网,无线

work on 影响;对……起作用;继续工作

wrist n. 手腕

III. Watching and Listening

Task One   Change the World, Love your Job

New Words

diffusion n. 扩散

fabrication n. 制造

instrument n. 仪器

manufacturing adj. 制造的

semiconductor n. [物] 半导体

logistics n. 后勤

planarization n.平面化

视频链接及文本

mentor n. 有经验的顾问　　　　　　　　module n. 模块；组件

rotate v 轮换；交替　　　　　　　　　　progression n 连续

Exercises

1. *Watch the video for the first time and choose the best answer to the following questions.*

　1) According to the video, the speaker _____.

　　　A. thought TI's technology manufacturing group rotation was a difficult way to find out

　　　B. has a lot of background in semiconductor

　　　C. doesn't like to experience the same roles to learn the business

　　　D. knew how to supply the engineering skills he learned in real world

　2) How does Kenny Palomino think about his first role? _____

　　　A. It is very difficult for him.

　　　B. It is easy for him.

　　　C. He can't play well in his first role.

　　　D. It is meaningful for him.

　3) Kenny Palomino's job was concentrated on _____.

　　　A. sharing his knowledge

　　　B. volunteering for support assignment

　　　C. the cost and water yield

　　　D. the process set and how to improve it in cost and wafer yield

　4) Kenny Palomino gained a lot of experience mainly because _____.

　　　A. he meets a lot of people from the different sides of the business

　　　B. the lab had a different equipment, different process and different metrics

　　　C. he gets to rotate from one module to another within the factory

　　　D. his working with many different people and different managers

　5) Kenny Palomino owned his career progression to _____.

　　　A. his hardworking

　　　B. the complete equipment

　　　C. working under different management styles

　　　D. others help

2. *Watch the video again and decide whether the following statements are true or false.*

　1) The speaker's first role was as a CMP or chemical plane engineer. (　)

　2) The speaker can share my knowledge and support tools that senior engineers are not familiar with. (　)

　3) The speaker has a lot of background in semiconductors. (　)

　4) The speaker's job is to work with a lot of different people, different managers so as to get a lot of experience. (　)

　5) The completed project made the speaker feel insecure. (　)

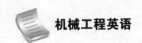

3. *Watch the video for the third time and fill in the following blanks.*

When I got my bachelor's in _____ engineering from the University of Texas at EI Paso, I knew I wanted to _____ my engineering skills that I learned in college to solve _____ problems, but I didn't know what that looked like. The technology _____ group rotation was a cool way to find out. I didn't have a lot of _____ in semiconductor, but I loved the idea of getting to _____ different roles and really learn the business. My first role was a CMP or _____ mechanical planarization engineer. That sounds _____, but what it means is I learned the _____ involved in wafer fabrication and I had _____ of a set of those processes.

4. *Share your opinions with your partners on the following topics for discussion.*

1) How do you like the day in the life of manufacturing engineer? Please summarize its features.

2) Can you use a few lines to list what's your understanding about manufacturing engineer? Please use an example to clarify your thoughts.

Task Two   Advanced Manufacturing

New Words

prototyping n. 原型机制造
solid adj. 固体的
thermoplastic adj. 热塑性的
polymers n. 聚合物
properties n. 特性
composite n. 复合材料
functional adj. 功能的

cross-section n.横断面
aerospace n.航空与航天
geometry n.几何形状
fiber n.光纤
transparent adj.透明的
extruder n.挤压机
composition n. 构图

视频链接及文本

Exercises

1. *Watch the video for the first time and choose the best answer to the following questions.*

1) We're interested in thermoplastic polymers because _____.

A. it is a rapid prototyping technique

B. it can be used with many different materials

C. it can be used to making blended composites and functional materials

D. it can be melted and deposited a layer at a time

2) Which of the following statement is false? _____

A. 3D printing could build complex structures.

B. We could build advanced design geometries previously impossible now.

C. It could build parts quickly.

D. We could send a CAD drawing to a 3D printer and it will arrive at a functional partwithin a day.

3) The word **address** here means _____.

A. speak   B. destination

C. call   D. work out

4) Wise composition control is _____.

A. a wide range of polymers and reinforcements

B. the control of the composition and functionality of a part

C. the control of the composition changing it on the fly is needed

D. both B and C

5) One of the drawbacks of conventional 3D printers is _____.

A. the high price   B. the complicated process

C. the long-time process   D. the build envelope

2. *Watch the video again and decide whether the following statements are true or false.*

1) 3D printing can be used in many of the materials that interest us in thermoplastic polymers because we can control their ultimate performance. (　)

2) Compounds are suitable for 3D printing because they melt and deposit one layer at a time. (　)

3) 3D printing can only build complex structures. (　)

4) The dynamic composition is built to capture the designer's imagination. (　)

5) We started a project to develop new thermoplastic compounds that could be used on 3D printers. (　)

3. *Watch the video for the third time and fill in the following blanks.*

When 3D printing was first _____ to the aerospace industry, we were eager to find ways to apply it. I understood 3D printing could build _____ structures. We realized that we could build _____ design geometries not _____ possible. We knew it could build _____ quickly, we could send a CAD _____ to a 3D printer, and we would arrive at a _____ part within a day. So early on we began a _____ to develop new thermoplastic _____ that could be used in a 3D printer and applied to our manufacturing _____.

4. *Share your opinions with your partners on the following topics for discussion.*

1) Do you know the features of 3D printing?

2) Can you give an example of 3D printing?

IV. Talking

Task One   Classical Sentences

Directions: *In this section, some popular sentences are supplied for you to read and to memorize. Then, you are required to simulate and produce your own sentences with reference to the structure.*

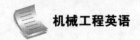

General Sentences

1. —What was the weather like yesterday?

   —Yesterday it rained all day.

   ——昨天天气怎样？

   ——昨天下了一天的雨。

2. —What will the weather be like tomorrow?

   —It's going to snow tomorrow.

   ——明天天气怎样？

   ——明天有雪。

3. Do you think it's going to rain tomorrow?

   你觉得明天会下雨吗？

4. I don't know whether it will rain or not.

   我不知道明天会不会下雨。

5. It'll probably clear up this afternoon.

   今天下午可能会放晴。

6. The days are getting hotter.

   天气在变暖。

7. What's the temperature today?

   今天多少度？

8. It's about twenty degrees centigrade this afternoon.

   今天下午大约 20 摄氏度。

9. There's a cool breeze this evening.

   今晚有股冷风。

10. Personally, I prefer winter weather.

    就我个人而言，我比较喜欢冬天。

11. What time are you going to get up tomorrow morning?

    明天早上你打算几点起床？

12. I'll probably wake up early and get up at 6:30.

    我可能会醒得比较早，大概六点半起床。

13. —What will you do then?

    —After I get dressed, I'll have breakfast.

    ——接着你做些什么呢？

    ——穿上衣服后我就去吃早饭。

14. What will you have for breakfast tomorrow morning?

    明天早餐你吃什么呢？

15. I'll probably have eggs and toast for breakfast.

    早餐我可能会吃吐司面包和鸡蛋。

16. After breakfast, I'll get ready to go to work.
    吃完早餐后我会准备一下去上班。

17. I get out of bed about 7 o'clock every morning.
    每天早上我 7:00 起床。

18. After getting up, I go into the bathroom and take a shower.
    起床后我会去浴室冲个澡。

19. Then, I brush my teeth and comb my hair.
    然后我刷牙梳头。

20. After brushing my teeth, I put on my clothes.
    我刷完牙后穿衣服。

21. After that, I go downstairs to the kitchen to have breakfast.
    然后,我下楼到厨房去吃早饭。

22. I'll leave the house at 8:00 and get to the office at 8:30.
    我 8 点离开家,8 点半到办公室。

23. I'll probably go out for lunch at about 12:30.
    我 12:30 左右去吃午饭。

24. I'll finish working at 5:30 and get home by 6 o'clock.
    我 5:30 下班,6:00 到家。

25. I'm always tired when I come home from work.
    下班以后我总是很累。

26. Are you going to have dinner at home tomorrow night?
    明天晚上你在家吃晚饭吗?

27. Do you think you'll go to the movies tomorrow night?
    明天晚上你会去看电影吗?

28. I'll probably stay home and watch television.
    我可能待在家看电视。

29. I'm not accustomed to going out after dark.
    我不习惯晚上出去。

30. When I get sleepy, I'll probably get ready for bed.
    当我感到困的时候,我就准备上床睡觉。

31. Do you think you'll be able to go to sleep right away?
    你觉得你现在能去睡觉吗?

32. What do you plan to do tomorrow?
    明天你打算干什么?

33. I doubt that I'll do anything tomorrow.
    明天恐怕我什么也不做。

34. I imagine I'll do some work instead of going to the movies.
    我想做点事,不想去看电影。

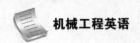

35. Will it be convenient for you to explain your plans to him?

你把你的计划跟他讲一下，方便吗？

36. What's your brother planning to do tomorrow?

你哥哥明天计划去哪儿？

37. It's difficult to make a decision without knowing all the facts.

不知道全部事实而做决定是很难的。

38. I'm hoping to spend a few days in the mountains.

我想在山上待几天。

39. Would you consider going north this summer?

你今年夏天想去北方吗？

40. If there's a chance you'll go, I'd like to go with you.

如果你有机会去的话，我想和你一起去。

41. After you think it over, please let me know what you decide.

在你仔细考虑之后，请告诉我你的决定。

42. Are you going to go anywhere this year?

今年你打算去哪儿？

43. If I have enough money, I'm going to take a trip abroad.

要是我有足够的钱，我打算出国旅行。

44. How are you going? Are you going by boat?

你打算怎样去？是不是乘船去？

45. It's faster to go by plane than by boat.

坐飞机比坐船快。

46. What's the quickest way to get there?

到那儿去最快的交通方式是什么？

47. Altogether it will take ten days to make the trip.

这次旅行总共要花十天时间。

48. It was a six-hour flight/journey/travel/voyage.

这是一次 6 小时的旅程。

49. I'm leaving tomorrow, but I haven't packed my suitcases yet.

我明天出发，可是我的箱子到现在还没有整理好。

50. I hope you have a good time on your trip.

祝你旅途愉快。

Specialized Sentences

1. When SI units are used, force is expressed in newtons (N) and area in square meters (m^2).

采用国际单位时，力用牛顿（N）表示，面积用平方米（m^2）表示。

2. When using USCS units, stress is customarily expressed in pounds per square inch(psi)

or kips per square inch(ksi).

当用美国单位制时,应力通常用磅/平方英寸(psi)或千磅/平方英寸(ksi)表示。

3. It is usually safe to assume that the formula $\sigma = F/A$ may be used with good accuracy at within the bar that is at least a distance d away from the ends, where d is the largest transverse dimension of the bar.

假设公式 $\sigma = F/A$ 能较好地适用于杆件内部离开两端距离至少不小于 d 的地方,这通常是安全的,此处 d 为杆的最大横向尺寸。

4. The principal requirement is that the deformation of the bar be uniform, which in turn requires that bar be prismatic, the loads act through the centrists of the cross sections, and the material be homogeneous (that is the same throughout all parts of the bar).

主要的要求是钢筋的变形是均匀的,这就要求钢筋是棱柱形的,载荷通过横截面的中心点的作用,并且材料是均匀的(即所有部分都相同)。

5. Capacity of a machine component is related to the most severe service condition it can sustain without a change which will prevent the component from continuing its intended function.

机器组件的承载能力与它所能承受的最恶劣的工作条件有关,而无须更改即可防止组件继续其预期功能。

6. Ultimate tensile strength is the maximum nominal stress which can be observed in the stress-strain diagram, which corresponds to the maximum nominal stress that the material can sustain, the ratio of maximum load to the original cross-sectional area.

最大拉伸强度是最大的名义应力,这可从应力—应变图上看出来,对应于材料承受的最大名义应力,是最大载荷对初始横截面积的比值。

7. Micro-cracks exist in the structure of this material which gives rise to high stress concentration during tensile loading, while in compressive loading, for geometrical reasons, these micro-cracks are ineffective.

微裂纹存在于这种材料结构中,拉伸时导致较大的应力集中,而压缩时,因几何原因,这些微裂纹失去作用。

8. A hollow shaft weighs considerably less than a solid shaft of comparable strength, but is somewhat more expensive.

在强度相当的情况下,空心轴的重量比实心轴的重量要轻许多,但其加工成本比实心轴要高。

9. Careful location of bearings can do much to control the size of shafts, as the loading is affected by the position of mountings.

由于零部件在轴上的安装位置对轴的受力有影响,仔细地布局好轴承位置对控制轴的尺寸大小很有帮助。

10. Often, the bore of such other components has to be chamfered to clear radii at the point where a shaft changes diameter.

通常,这类零件的孔需加工成锥面,以跳过轴颈变化处的过渡半径。

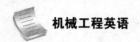

11. The diameters of shafts made compatible with metric-sized bores of mechanical components (such as antifriction bearings) are specified in millimeters.

    与公制尺寸的机械零件(如抗磨轴承)镗孔兼容的轴的直径以毫米为单位。

12. One rule of thumb is to restrict the torsional deflection to one degree in a length equal to 20 diameters.

    一种经验方法是,在每20倍直径的轴的长度上,扭转挠度不超过1度。

13. When two gears are meshing, the two pitch circles must be exactly tangent if the gears are to function properly.

    两个齿轮啮合时,若要求它们正常运转,两个齿轮的节圆必须相切。

14. Pressure angle: the angle between the line of action and a line perpendicular to the centerlines of the two gears in mesh.

    压力角:作用线与垂直于两啮合齿轮中心线的直线间的夹角。

15. The metric system for spur-gear calculations involves more than changing inches to millimeters.

    采用米制进行齿轮的计算,所做的工作不仅仅是将英尺换算成毫米。

16. The module is found by dividing the pitch-circle diameter by the number of teeth.

    该模块是通过将螺距圆直径除以齿数来实现的。

17. The module is thus a dimension, expressed in millimeters, equal to the reciprocal of the diametric pitch.

    因此,模数是一个线性尺度,用毫米表示,它等于径向间距的倒数。

18. Designer should carefully determine horizontal force components, since these problems designing the follower assembly guide.

    设计师应仔细分析水平力量,因为它们会使动件的导向出现问题。

19. Again, forces need to be determined and dimensions chosen so as to avoid excessive component sizes.

    同样,需要分析受力,选择尺寸,以避免零件尺寸过大。

20. Some of them make it more comfortable or better looking, but most of them are to make it run.

    有些零件是让汽车更舒适、更美观的,而大多数零件是用来使它运动的。

21. Typical materials that can be cast with this process are aluminum, iron, steel, nickel alloys, copper alloys.

    这种工艺所用的典型材料有:铝、铁、钢、镍合金、铜合金。

22. This electrode may be used for poor welding condition.

    焊接条件差可使用这种焊条。

23. However, because of the heavy coating, there are about half as many rods per pound.

    然而,因涂层较重,每磅焊条数大约是其他涂层焊条的一半。

24. However, it should be fully understood that any increase of strength over that of pure iron can be obtained only at the expense of some loss of ductility, and the final choice is always a compromise of some degree.

然而,必须明白,要使其强度高于纯铁的强度只有在损失塑性的情况下才能实现,最终是在塑性和强度之间形成某种折中。

25. Steels with approximately 6 to 25 points of carbon(0. 06% ~ 0. 25%) are rated as low carbon steels and are rarely hardened by heat treatment.

碳的质量分数为 0.06% ~0.25% 的钢称为低碳钢,它们很难通过热处理淬硬。

26. Even plain carbon steels are alloys of at least iron, carbon ,and manganese, but the term alloy steel refers to steels containing elements other than these in controlled quantities greater than impurity concentration or ,in the case of manganese, greater than 1.5%.

普通碳素钢也是铁、碳和锰的合金,但合金钢中除了这些元素外,其他元素含量大于普通碳素结构钢的杂质含量,如锰的含量要大于 1.5%。

27. Any alloy generally shifts the transformation curves to the right, which with air cooling results in finer pearlite than would be formed in plain carbon steel.

合金通常使转变曲线右移,通过空气冷却,其珠光体比普通碳钢中的珠光体细。

28. The carbon content may vary from very low to very high, but for most steels it is in the medium range that effective heat treatment may be employed for property improvement at minimum costs.

碳含量可以从很低变到很高,但大多数为中碳钢,可用最小成本热处理来有效改善性能。

29. Plain low carbon steels, when fully annealed, are soft and relatively weak, offering little resistance to cutting, but usually having sufficient ductility and toughness that a cut chip tends to pull and tear the surface from which it is removed, leaving a comparatively poor quality surface, which results in a poor mach-inability rating.

当完全退火时,普通低碳钢硬度较低,强度较小,对切削的阻力较小,但通常由于塑性和韧性太大以致切削离开工件表面时会划伤表面,工件表面质量比较差,导致较差的机能等级。

30. The mach-inability of many of the higher plain carbon and most of the alloy steels can usually be greatly improve by annealing, as they are often too hard and strong to be easily cut at any but their softest condition.

许多高碳钢和大多数合金钢的机能等级通常可以通过退火改善,因为除在最软条件下,它们的硬度和强度太高而不易切削。

31. Redundant robots are those that have more degrees of freedom than those required to perform a given task.

冗余机器人是指比执行给定任务所需的自由度更高的机器人。

32. It is known that its restricted manageability results in a reduction of the work space due to the mechanical limitations of the joints and to the presence of obstacles in that space.

众所周知,由于关节的机械限制和障碍物在那个空间中的存在,有限的可操作性会导致工作空间减少。

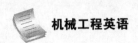

33. It's by no means an easy job to produce such a tool without the help of modern technology.

没有现代技术的帮助,想生产这样的工具绝非易事。

34. Fig. 2 is a schematic diagram of the SCARA-type redundant manipulator showing its redundancy in its rotational motion

图 2 是 SCARA 型冗余机械手的示意图,显示了它在旋转运动中的冗余性。

35. Now we make the corresponding calculations for the design of a kinematic model of the manipulator.

本文对机械手的运动学模型的设计进行了相应的计算。

36. To get the kinematic model the standard method of Denavit-Hartenberg has been considered, whose parameters are indicated in Table 1.

为了得到运动学模型,可以使用 Denavit-Hartenberg 的标准方法,具体参数见表 1。

37. Then, using the homogeneous transformations given we get the direct kinematic model indicated by matrix equation.

然后,利用齐次变换,得到由矩阵方程表示的直接运动学模型。

38. EMG sensors detect these signals and translate them into control signals that cause the mechanical hand to open and close.

EMG 感受器检测到这些信号,把它们转换成引起机械手张开或握住的控制信号。

39. That is case, for simplicity, the condition $\theta 4 = \theta 3$ is set.

为了简单起见,将 $\theta 4 = \theta 3$ 的条件设定好。

40. In accordance with this, and after the adequate simplifications, we get the inverse kinematic model expressed by Eqs.

在此基础上,经过充分的简化,得到了用 Eqs 表示的逆运动学模型。

41. Keeping in mind the characteristics of the manipulator presented so far, we get its dynamic model.

针对目前提出的机械手的特点,我们得出了机械手的动力学模型。

42. Getting the inverse kinematics of a redundant robot requires looking at different methods and selecting the most adequate one according to the considerations of the model.

求冗余度机器人的逆运动学需要根据模型考虑不同的方法,并选择最合适的方法。

43. In this work the Lagrange-Euler formulation that is based on the principle of the conservation of energy is used.

在这项工作中,使用了基于能量守恒原理的拉格朗日—欧拉公式。

44. Robot hand mimics not just the movement, but also the means of achieving it.

机械手不仅模仿手的运动,还能模仿完成这个动作的方法。

45. Fig. 3 shows a schematic of a servo motor coupled with a robotized manipulator as the load.

图 3 给出了一个伺服电机的原理图,以机械手作为负载。

46. The demonstration checked on the avionics, wiring, gear boxes, cockpit seat and electronics, among other things.

这次展示包括航空电子设备、配线、齿轮箱、座椅和座舱电子设备以及其他设备。

47. The feedback control circuit that converts an input signal of the PWM (Pulse-Width Modulation) type to voltage, comparing it with the fed back position.

该反馈控制电路将 PWM(脉宽调制)型输入信号转换为电压,并将输入信号与反馈位置进行比较。

48. Fig. 4 shows a block diagram of an analogic servo motor connected with a load consisting of a robot.

图 4 给出了一个与机器人构成的负载连接的模拟伺服电机的框图。

49. The actuators considered in this study correspond to analogic servo motors.

本研究中所考虑的执行器对应于模拟伺服电机。

50. One of the advantages of the sliding mode control is its in-variance when facing parametric uncertainties and external perturbations.

滑模控制的优点之一是它在面对参数不确定性和外部干扰时具有不变性。

Task Two Sample Dialogue

Directions: *In this dialogue, you are going to read several times the following sample dialogue about the relevant topic. Please pay special attention to five Cs (culture, context, coherence, cohesion and critique) in the dialogue and get ready for a smooth communication in the coming task.*

(*A Dialogue between a welding engineer and a modeling engineer*)

Walter
(**welding engineer**): Hi, Michael. I know you are doing modeling work for welding but I am not sure how this can help people like me. Could you explain it to me?

Michael
(**modeling engineer**): Yes, Walter. As we all know, various welding methods have been widely adopted to join materials together. There are all kinds of welding-related issues. Let me find an example to show how the modeling can help.

Walter: How about welding-induced distortion? I was doing an arc-welding yesterday. It was a simple butt joint. I clamped two plates on the anvil as I learned in my training. After the welded plates cooled down, I took them from the anvil and found they are no longer flat.

Michael: As you know, welding involves intense heat input in a localized area. The heated material will expand its volume. However, this volume

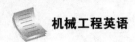

expansion is resisted by the neighboring materials. The interaction between the heated material and the neighboring material results in stress. A similar scenario occurs when the welded parts cool down, except the neighboring material is now resisting volume shrinkage. It is the stress that causes the distortion.

Walter: Yes, I have heard something like this before but I don't know if I really understand it. I can see distortions, but I cannot see stress.

Michael: OK, that is something I can help you. Temperature and stress are the basic physical quantities we are interested. In modeling, we try to predict them.

Walter: How do you do that?

Michael: The change of temperature and stress in welding process is controlled by some physical laws which we call governing equations. To predict temperature and stress, we solve the equations. To give you an analogy, think of this: You throw a piece of stone in the air. The stone flies along a trajectory, going up first then falling down. This trajectory can be calculated based on the fundamental physics.

Walter: Are you talking about Newton's law? I learned this in school.

Michael: Yes, it is called Newton's law. For temperature and stress, we have similar laws.

Walter: Then, you will have to solve them. It is not easy to get the solution, is it? I remember it took me a while to calculate how far the stone can fly.

Michael: You are right that it is not easy to solve them through hand calculation. Nowadays, we use computer and engineering software to solve the problem.

Walter: Could you tell me more? I am curious to know.

Michael: Basically, you have to let the software know what problem you want to investigate. This includes a lot of details such as the geometry and material of the welding, the welding conditions, and the issues you are interested in.

Walter: So, it sounds like I have to be familiar with the software. This will require certain training and practice. Right?

Michael: Yes, but I have some good news for you. In order to help welding engineers, EWI has developed a welding analysis tool called EWI Weld Predictor. The tool is essentially the customization of the analysis of common arc welding processes, such as GMAW, GTAW, etc. It can analyze the welding of a wide range of plate and pipe joints. Welding engineers can input the welding parameters they are interested through a

	Graphic User Interface and launch analysis to predict the residual stress, distortion, and micro-structure. After the analysis is finished, a report will be generated to summarize the results. The analysis tool is installed on EWI's members-only website. It is free for EWI members to use.
Walter:	Wow. This seems to be very attractive. Since our company is a member, I'd like to give a try. Do you offer any training on it?
Michael:	The use of this analysis tool is very straightforward. Simply follow the web pages and the instructions. You will find no training is needed. I will set up an account for you and send you the login information. At this point, I would suggest you to login to Weld Predictor and try to analyze a few welding processes. Then, based on the analysis, we can have further discussions on the distortion issues you are interested in.
Walter:	Great. Thank you very much. I definitely will try it and get back to you.

Task Three   Simulation and Reproduction

Directions: *The class will be divided into three major groups, each of which will be assigned a topic. In each group, some students may be the teacher, while others may be students. In the process of discussion, please observe the principles of cooperation, politeness and choice of words. One of the groups will be chosen to demonstrate the discussion to the class.*

1) advanced manufacturing technologies in our daily life
2) a funny story related to manufacturing productions in my childhood
3) the importance of learning advanced manufacturing technologies

Task Four   Discussion and Debate

Directions: *The class will be divided into two groups. Please choose your stand in regard to the following controversy and support your opinions with scientific evidences. Please refer to the specialized terms and classical sentences in the previous parts of this unit.*

The debate within manufacturing about whether technology will completely replace people is interesting. Technology is changing the workforce, and it has eliminated low-skilled manufacturing jobs in the past. Do you agree?

V. After-class Exercises

1. *Match the English words in Column A with the Chinese meaning in Column B.*

A	B
1) mechanical drawing	A) 构图

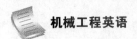

2) probability   B) 半导体

3) cams   C) 主机

4) mainframe   D) 概率

5) industrial robot   E) 工业机器人

6) module   F) 航空

7) semiconductor   G) 凸轮

8) fiber   H) 光纤

9) composition   I) 机械制图

10) aerospace   J) 模块

2. *Fill in the following blanks with the words or phrases in the word bank. Change the forms if it's necessary.*

autonomously	component	manufacturing	automation	smart
raw material	pneumatic	internet	artificial	manually

1) The history of automation in the manufacturing industry can betraced back to the early use of basic _____ systems.

2) Many operations in the _____ industry have become automated, increasing production as well as labor.

3) Great advances have been made in the automation of the various activities formerly carried out _____.

4) The automobile industry is another example of _____, with computers eventually taking over a lot of manual activities.

5) Japan was in the forefront of developing _____ for use in industrial manufacturing automation.

6) Manufacturing engineering are the steps through which _____ are transformed into a final product.

7) _____ manufacturing enables all information about the manufacturing process to be available when it is needed, where it is needed.

8) Advanced robots operate _____ and can communicate with manufacturing systems.

9) Open communication between manufacturing devices and networks can also be achieved through _____ connectivity.

10) These robots are able to complete work beyond what they were initially to do and have _____ intelligence.

3. *Translate the following sentences into English.*

1) 设计者应仔细分析水平力分量,因为它会使从动件的导向出现问题。

    _____

    _____

2) 机器零件的承载能力是承受最恶劣的工作条件而不产生影响其功能的能力。

3)有些零件是让汽车更舒适、更美观的,而大多数零件是用来使它运动的。

4)这种工艺所用的典型材料有:铝、铁、钢、镍合金、铜合金。

5)碳含量可从很低变到很高,但大多数为中碳钢,可用最小成本热处理来有效改善其性能。

4. *Please write an essay of about 120 words on the topic:* **Application of 3D printing in our life.** *Some specific examples will be highly appreciated and you have to watch out the spelling of some specialized terms you have learnt in this unit.*

VI. Additional Reading

The Assembly Line for Car Production

In a newly built factory in 1913, Ford Motor Company introduced the assembly line for car production. Prior to this, single cars were built by a number of skilled and unskilled workers, in an old factory. So the assembly line can be considered one of the first forms of automation in the manufacturing industry. It certainly boosted Ford Motor's production rates, as well as their profits, but he was very good to his employees, giving them a rate of pay over and above other industries in the area. Their pay even allowed them to own one of the cars they produced, which was unheard of in the industry. Ford's assembly line and mass production was the first in the world; cutting the car assembly time from one car every twelve hours to a car every one and a half hours.

Japan was in the forefront of developing components for use in industrial manufacturing automation. During the 1930s one of their forward looking companies developed a highly accurate electrical timer, along with the first micro-switch and protective relays. All of these were immediately used in industry. At about this time, the rest of the world were beginning to see the advantages of automation, and a lot of research and development was taking place, with the major component being a solid state proximity switch. During the Second World War between 1939 and 1945, automation continued especially in tanks, warships, fighter airplanes, and landing craft used to get the soldiers ashore from troop carriers.

Japan officially surrendered in 1945; the US and Allied Forces occupied Japan from then until 1952. An industrial rebuilding program, assisted mainly by the US was immediately started. This meant they were using new technology, including the latest automation that was far superior to the rest of the world who were mostly still manufacturing goods using old fashioned methods. Japan was soon to become a world leader in automation, especially in the automobile industry. Nissan, Toyota, and Honda produced thousands of new high quality, reliable, modern cars. These had standard accessories that most other car manufacturers classed as extras. They were able to do this because of the money saved using automation technology. These high standards, coupled with realistic prices that could not be bettered by other car manufacturers, ensured their success in this industry. I was engineer on a copper smelter Zambia in the 1970s and bought an old British Wolseley car for getting back and forth to work. It blew a cylinder head gasket and I tried everywhere to get one, but it was pretty old and I had no success. I wrote to a dealer in the UK for spares, but in between times I was told that a Toyota four-cylinder engine was the same as the Wolseley, and these gaskets were available from the local Toyota garage! It seems that Toyota bought the design plates for the Wolseley engine, and fashioned one of their engines on them; part of their rebuilding their industry in the 1950s.

Tool making and precision machining are the need of the day, but recent advances in technology demand ever higher precision and accuracy in order to reach the desired limit. This article will serve as a guide for tool making and precision machining articles at Bright Hub. Technology has been improving the machines and gadgets we use in our day to day life. Starting with the huge gramophones in the olden days from which we used to hear music, today's technology has shrunk the gramophone to an iPod that fits in the palm of your hand. The technological advances have created a need to make things better and better things to be their best. This metamorphism of technology has led to the evolution of machining, pushing it from the level of micro accuracy to the level of nana accuracy. New generation machines are no longer manually controlled and operated; the computer has taken control over them now. New generation machines are now capable of achieving accuracy levels which were only talked about a decade ago.

A part or component is manufactured from its blank stage on various machines to attain its final shape. Machines normally remove material from a blank in order to provide the required shape. Each kind of machine removes material in a specific way to the required dimensions as controlled by the operator of the machine. Every type of machine requires its own unique type of tooling that aids in the material removal.

1. *Read the passage quickly by using the skills of skimming and scanning, and choose the best answer to the following questions.*

1) How was the individual car manufactured before the introduction of the car line? _____
   A. made by skilled workers
   B. made by unskilled workers
   C. made in a factory
   D. built by skilled and unskilled workers in an old factory

2) Why did the assembly line increase Ford's cycling productivity? _____
   A. Because Ford was good to his workers.
   B. Because Ford was very strict.
   C. Because of automation in the manufacturing industry.
   D. Because of high rate of pay.

3) How long does it take ford to make one car after the assembly line? _____
   A. 3 hours   B. 6 hours
   C. 12 hours   D. 1.5 hours

4) Where does Japan stand in developing components for industrial manufacturing automation? _____
   A. backward status   B. leading position
   C. high-end status   D. unknown

5) A Japanese company developed a highly accurate electronic timer, as well as the first miniature switches and protection relays which were applied to _____?
   A. automobile making   B. agriculture
   C. industry   D. service

6) What is the most remarkable aspect of the development of automation during the Second World War between 1939 and 1945? _____
   A. landing craft   B. automobile making
   C. industry   D. ianding craft

7) When did Japan formally surrender? _____
   A. 1939   B. 1945
   C. 1952   D. 1954

8) What does it mean to start a major industrial reconstruction program in the United States? _____

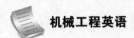
A. The United States is ahead of the rest.

B. The United States uses the latest technology.

C. The United States began industrial reform.

D. The United States defeated the Japanese.

9) What does the transformation of technology lead to? _____

A. development of machining

B. the advent of nanotechnology

C. The phonograph becomes smaller.

D. Things get better. Things get best.

10) What will the next generation of machines achieve? _____

A. a level that people can't imagine

B. the parts reach their final shape

C. Each type of machine needs its own unique tool to help.

D. The level of precision that was mentioned only 10 years ago.

2. *In this part, the students are required to make an oral presentation on either of the following topics.*

1) the brief introduction of advanced manufacturing technologies

2) the application to engineering

习题答案

Unit Four　Electromechanical Integrated Design

I. Pre-class Activity

Directions: *Please read the general introduction about* **Michael Faraday** *and tell something more about the great scientist to your classmates.*

Michael Faraday (1791—1867) was a British scientist who contributed to the study of electromagnetism and electrochemistry. His main discoveries include the principles underlying electromagnetic induction, diamagnetism and electrolysis.

Although Faraday received little formal education, he was one of the most influential scientists in history. It was by his research on the magnetic field around a conductor carrying a direct current that Faraday established the basis for the concept of the electromagnetic field in physics. Faraday also established that magnetism could affect rays of light and that there was an underlying relationship between the two phenomena. He similarly discovered the principles of electromagnetic induction and diamagnetism, and the laws of electrolysis. His inventions of electromagnetic rotary devices formed the foundation of electric motor technology, and it was largely due to his efforts that electricity became practical for use in technology.

Michael Faraday invented the first electric motor in 1821. Ten years later the first electric generator was invented, again by Michael Faraday. This generator consisted of a magnet passing through a coil of wire and inducing current that was measured by a galvanometer. Faraday's research and experiments into electricity are the basis of most of modern electromechanical principles known today.

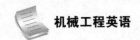

II. Specialized Terms

Directions：*Please memorize the following specialized terms before the class so that you will be able to cope with the coming tasks better.*

actuate vt. 使动作;开动

actuator n.［机］促动器;［电］(电磁铁)螺线管

algorithm n. 运算法则;演算法;计算程序

alternator n. 交流发电机

ASME abbr. American Society of Mechanical Engineers 美国机械工程师协会

automatic transmission system 自动发送系统

automotive adj. 自动的;汽车的

control loop 控制回路,控制环,操纵系统,驾驶系统

controller n. 控制者;(机器的)控制器

converter n. 变换器;换流器,变压器,变频器

rotary converter 同步换流机,旋转变流器,电动机发电机组

crossbar switch 纵横开关

current n. 电流

cybernetics n. 控制论

device n. 装置,设备;方法;策略;手段

diamagnetism n.反磁性,逆磁性,反磁性学

dielectric n. 电介质,绝缘体; adj. 非传导性的

direct current 直流电

disturbance n. 干扰,打扰

electric generator 发电机

electric motor 电动机

electric typewriters 电动打字机

electrical engineering 电机工程,电工

electrical output 电输出

electrical signal 电信号

electrify vt. 使电气化;使充电,使通电

electrochemistry n. 电化学

electromagnetic adj. 电磁的

electrohydraulic forming 电液成形

electromagnetic induction 电磁感应

electromagnetic pump 电磁泵

electromagnetic transducer 电磁式传感器

electromagnetism n. 电磁,电磁学

electromechanical brake 电闸

electromechanical pickup 机电传感器

electromotive force 电动势

electron beam welding 电子束焊

electrolysis n. 电解,电蚀

electromotive adj. 电测的

electromotive force 电动势

electronics engineering 电子工程

engine n. 发动机,引擎;vt. 给……安装发动机

expert system 专家系统

generator n. 发电机,发生器;电力公司

high-voltage 高压的

IEEE abbr. Institute of Electrical and Electronic Engineers 电气和电子工程师协会

integrated circuit 集成电路

integration n. 结合;整合;一体化

interaction n. 一起活动;互相影响;互动

intervention n. 介入,干涉,干预;调解,排解

isolated circuit voltage 隔离电路电压

regulator n. 调整器,校准器,调节器

linear regulator 线性调节器

linkage n. 联系,连锁,联动,连接;链系,

联动装置

magnet n.磁铁,磁石;[物]磁体

magnetic field n. 磁场

manually-operated 人工操作的

mechanical engineering n. 机械工程(学)

mechanics n. 力学;机械学

mechatronics n. 机电一体化

mechatronic module 机电模块

microcontroller n. 微控制器

microelectronic adj. 微电子(学)的

microprocessor n. 微处理器

module n. 模块;组件

motor n. 马达,发动机;汽车

motor-generator 电机发电机

multidisciplinary adj. 包括各种学科的;多
学科;多科目

numerical control machine 数控机床

optoelectronic adj. 光电子的

panel switch 面板开关

piezoelectric adj. 压电的

piloted valve 导向阀,引导阀

planar linkage 平面连杆机构

polytechnic adj. 综合技术的;各种工艺的

power n. [机]动力,功率;力量 vt. 运转;
用发动机发动

power conversion 能量[功率]变换

power-assisted 动力辅助的

precision n. 精确度,准确(性) adj. 精确
的,准确的

pump n. 抽水机;打气筒;泵 v. 用泵输送

rectifier n. 整流器

rotary device 旋转设备

sensor n. 传感器,灵敏元件

sequential control 顺序控制(方式)

servo-mechanics 伺服机构,伺服机件

simulate vt. 模仿;模拟

solenoid valve 电磁阀,螺线管操纵阀

Strowger switch 史端桥开关

subsystem n. 子系统,分系统

supercharger n. 增压器;增压机

switchgear n. 接电装置,开关设备

synergistic adj.增效的,协作的,互相作用
(促进)的

telecommunications engineering 电信工程

telegraph signal 电报信号

telegraphy n. 电信技术;超感

thermostat n. 恒温(调节)器

turbine n. 涡轮机;透平机

turbo n. 涡轮(发动机),增压涡轮;同
"turbo supercharger"

transformer n. 变压器

transistor n. 晶体管;晶体管收音机,半导
体收音机

transmitter n. 传送者;传达者;发射机;发
报机

well-oiled adj. 运转顺畅的,运行良好的

wind turbine 风轮机,风力涡轮机

III. Watching and Listening

Task One   The Greatest Machine in History(Ⅰ)

New Words

prototype n. 原型,雏形,蓝本   v. 作为原
型,提供蓝本

affinity n. 密切关系,姻亲
关系;(男女之间的)吸引

视频链接及文本

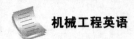
机械工程英语

力;类似,近似

unkempt adj. (头发)蓬乱的

aristocracy n. 贵族;贵族品质;贵族统治的
国家;上层社会

intelligentsia n. 知识分子;知识界

soiree n. 晚会,黄昏时的聚会

mathematician n. 数学家

nerd n. 令人讨厌的人,卑微的人

bid vt. 恳求;命令;说(问候话)

monstrosity n. 庞然大物

analytical engine 解析机,分析机(早期的
机械通用计算机)

complexity n. 复杂性

regularity n. 规则性,规律性

microscope n. 显微镜

punch card 穿孔卡片

reader n. 阅读器,读取器

Jacquard loom 提花织机

repurpose vt. 赋予新的用途

revolutionize vt. 彻底改革;使革命化

dynamo n. 发电机

proven adj. 经过验证或证实的

cog n. 齿轮 vt. 以雄榫连接

locomotive n. 火车头;机车

memory n. 存储器,内存

decimal adj. 十进位的,小数的

binary adj. 双重的;二元的

arithmetic n. 算术,计算;算法

feat n. 卓绝的手艺,技术,本领

silicon chip 硅片

Exercises

1. *Watch the video for the first time and choose the best answer to the following questions.*

1) About Charles Babbage, which of the following statements is NOT true? _____

A. He was born at the end of the 18th century.

B. He was a very wealthy man.

C. He was a famous mathematician in his time.

D. He tried to make mechanical computing devices.

2) If you were _____, you would most probably have been invited to Babbage's house
for a party.

A. an intellectual   B. a wealthy man

C. a mechanic   D. a movie star

3) The machine designed by Charles Babbage was _____.

A. tiny   B. gigantic

C. a steam locomotive   D. ridiculous

4) The machine was said to be a computer because it _____.

A. had the memory   B. had a CPU

C. could make the "if then" decision   D. had all of the above features

5) The technologies that Babbage used in designing this machine do NOT include _____.

A. cogs   B. the steam

C. the dynamo   D. punch card

2. *Watch the video again and decide whether the following statements are true or false.*

1) The first computer was designed in the 1930s and 1940s. (   )

2) Although he held the post that Newton once held at Cambridge, Charles Babbage was less famous than Newton. (　　)

3) Babbage never made any of the mechanical devices he designed because he thought they were too complicated. (　　)

4) The machine Babbage designed was actually a simple calculator. (　　)

5) The Jacquard loom was also designed by Charles Babbage. (　　)

3. *Watch the video for the third time and fill in the following blanks.*

     And I'm going to take you through the _____ of the machine—that's why it's computer architecture. Firstly, let's talk about the _____. The memory is very like the memory of a computer today, except it was all made out of _____, stacks and stacks of metal _____. And they've got _____ on them. It's a _____ machine. And he thought about using binary. The problem with using binary is that the machine would have been so tall, it would have been _____. Also, this machine has its CPU, the chip. And the CPU is very big and completely _____. The CPU could do the four fundamental functions of arithmetic—addition, multiplication, subtraction, division. But it could also do something that a does and a _____ doesn't: this machine could look at its own _____ memory and make a decision.

4. *Share your opinions with your partners on the following topics for discussion.*

1) In what ways is the machine designed by Charles Babbage different from the computer of today? In what ways are they similar?

2) What is mechatronics? Can you explain it in your own words based on your professional knowledge?

Task Two　The Greatest Machine in History（Ⅱ）

New Words

accessory n. 附件 adj. 附加的;附属的

difference engine 差分机

obsessed adj. 着迷的

graphics n. 图表;图样

plotter n. 绘图机;标图员,制图者

plot v. 以图表画出,制图

inherit vt. & vi. 继承;vt. 经遗传获得(品质、身体特征等),继任

function n. 函数

leap v. 跳;(使)跳跃;n. 跳跃,飞跃

nanotechnology n. 纳米技术,毫微技术;纤技术

视频链接及文本

thesis n. 论点,论题;命题

RAM　abbr. Random Access Memory 随机处理器

archive v. 存档　n. 档案文件

curator n. 馆长;监护人;管理者

simulation n. 模仿,模拟

humongous adj.<俚>极大的,具大无比的

fiddle（with）vi. 神经质地摆弄手指或手

Exercises

1. *Watch the video for the first time and choose the best answer to the following questions.*

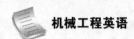

1) _____ was not mentioned by the speaker as one of the accessories the machine has.

    A. a bell to ring   B. a printer

    C. a plotter   D. a big piece of paper

2) Who built the Difference Engine No. 2 which has the printing mechanism? _____

    A. Charles Babbage   B. Ada Lovelace

    C. Alan Turning   D. The Science Museum

3) Who was the only person in Babbage's life that could understand the future of his machine? _____

    A. Ada Lovelace   B. Alan Turning

    C. Lord Byron   D. The Duke of Wellington

4) It's _____ that underlie(s) a phone or computer or any other computing device.

    A. texts   B. graphics

    C. numbers   D. music

5) The machine designed by Babbage _____.

    A. is gigantic and fast

    B. is gigantic and slow

    C. runs as fast as today's computers

    D. less capable than today's computers

2. *Watch the video again and decide whether the following statements are true or false.*

1) The mechanical printer Babbage designed can print numbers and pictures. (　)

2) Ada Lovelace's mother made Ada get mathematic training because she had shown great talents in mathematics. (　)

3) Historically, Ada Lovelace isn't the first programmer, but she did something more amazing. (　)

4) It was Alan Turning who laid down the mathematical foundation for computer science. (　)

5) When Babbage's machine is built, we're able to observe how it works and thus to understand how a computer works just because it runs rather slowly. (　)

3. *Watch the video for the third time and fill in the following blanks.*

    A hundred years later, a guy named Alan Turning came along _____, and in 1936 he _____ the computer all over again. Of course, Babbage's machine was entirely _____ while Turning's machine was entirely _____. Both of them were coming from a _____ perspective, but Turning told us something very important. He laid down the mathematical foundation for _____, and said, "It doesn't matter how _____ you." It doesn't matter if your computer is mechanical or _____, or perhaps in the future, or mechanical again, once we get into _____. We could go back to Babbage's machine and just make it _____. All those things are computers.

4. *Share your opinions with your partners on the following topics for discussion.*

1）What has electro-mechanics brought to our life?

2）With the development of mechatronical technologies, what will our life be like in 50 year?

IV. Talking

Task One　Classical Sentences

Directions: *In this section, some popular sentences are supplied for you to read and to memorize. Then, you are required to simulate and produce your own sentences with reference to the structure.*

General Sentences

1. Do you really want to know what I think?
   你真想知道我在想什么吗？

2. Please give me your frank opinion.
   请告诉我你真实的想法。

3. Of course I want to know what your opinion is.
   当然,我很想知道你的看法。

4. What do you think? Is that right?
   你认为怎样？是对的吗？

5. Certainly. You're absolutely right about that.
   当然,你完全正确。

6. I think you're mistaken about that.
   我想你弄错了。

7. I'm anxious to know what your decision is.
   我很想知道你的决定。

8. That's a good/great/fantastic/excellent idea.
   这个想法很好。

9. In my opinion, that's an excellent idea.
   我认为这是个好主意。

10. I'm confident you've made the right choice.
    我相信你做了个正确的决定。

11. I want to persuade you to change your mind.
    我想劝你改变主意。

12. Will you accept my advice?
    你会接受我的建议吗？

13. He didn't want to say anything to influence my decision.

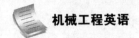

他不想说任何话来影响我的决定。

14. She refuses to make up her mind.

她不肯下决心。

15. I assume you've decided against buying a new car.

我想你已经决定不买新车了吧。

16. It took him a long time to make up his mind.

他用了很久才下定决心。

17. You have your point of view, and I have mine.

你有你的观点,我也有我的想法。

18. You approach it in a different way from mine.

你处理这件事的方式和我不一样。

19. I won't argue with you, but I think you're being unfair.

我不想和你争辩,但是我认为你是不公平的。

20. That's a liberal point of view.

那是一种自由主义的观点。

21. He seems to have a lot of strange ideas.

他好像有很多奇怪的想法。

22. I don't see any point in discussing the question any further.

我看不出有进一步讨论这个问题的必要。

23. What alternatives do I have?

我还有什么方法?

24. Everyone is entitled to his own opinion.

每个人都有自己的观点。

25. She doesn't like anything I do or say.

无论我做什么说什么,她都不喜欢。

26. There are always two sides to everything.

任何事物都有两面性。

27. We have opposite views on this.

关于这个问题,我们有不同的观点。

28. Please forgive me. I didn't mean to start an argument.

请原谅,我并不想引起争论。

29. I must know your opinion. Do you agree with me?

我必须了解你的想法。你同意吗?

30. What points are you trying to make?

你想表达什么观点?

31. Our views are not so far apart, after all.

毕竟我们的意见没有多大分歧。

32. We should be able to resolve our differences.

我们应该能解决我们的分歧。

33. If you want my advice, I don't think you should go.
   如果你征求我的意见,我认为你不应该去。

34. I suggest that you tear up the letter and start over again.
   我建议你把信撕掉,再重新写一遍。

35. It's only a piece of suggestion, and you can do what you want to.
   这只是一个建议,你可以做你想做的。

36. Let me give you a little fatherly advice.
   让我给你提点建议。

37. If you don't like it, I wish you would say so.
   如果你不喜欢,我希望你能说出来。

38. Please don't take offense. I only wanted to tell you what I think.
   请别生气,我只是想告诉你我的想法。

39. My feeling is that you ought to stay home tonight.
   我觉得你今晚应该待在家里。

40. It's none of my business, but I think you ought to work harder.
   这不关我的事,但我认为你应该更努力地工作。

41. In general, my reaction is favorable.
   总的来说,我是赞成的。

42. If you don't take my advice, you'll be sorry.
   如果你不听我的劝告,你会后悔的。

43. I've always tried not to interfere in your affairs.
   我总是尽量不干涉你的事情。

44. I'm old enough to make up my own mind.
   我已经大了,可以自己做决定了。

45. Thanks for the advice, but this is something I have to figure out myself.
   谢谢你的建议,但我必须自己解决问题。

46. He won't pay attention to anybody. You're just wasting your breath.
   他谁的建议都不会听,你们是在白费口舌。

47. You can go whenever you wish.
   你愿意什么时候去就什么时候去。

48. We're willing to accept your plan.
   我们愿意接受你的计划。

49. He knows it's inconvenient, but he wants to go anyway.
   他知道不方便,但他无论如何都想走。

50. He insists that it doesn't make any difference to him.
   他坚持说这对他没有任何影响。

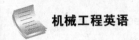

Specialized Sentences

1. An industrial robot is a prime example of a mechatronics system.

   工业机器人是机电一体化系统的最好例子。

2. Engineering cybernetics deals with the question of control engineering of mechatronic systems.

   工程控制论研究的是机电系统控制工程问题。

3. This device is used to control or regulate a system.

   这个装置被用来控制或调节系统。

4. Through collaboration, the mechatronic modules perform the production goals and inherit flexible and agile manufacturing properties in the production scheme.

   通过协作,机电模块实现了生产目标,并在生产方案中继承了灵活敏捷的制造特性。

5. Mechatronics is a multidisciplinary field of engineering that includes a combination of mechanical engineering, robotics, electronics, computer engineering, telecommunications engineering, systems engineering and control engineering.

   机电一体化是一个多学科的工程领域,包括机械工程、机器人、电子、计算机工程、电信工程、系统工程和控制工程。

6. As technology advances, the subfields of engineering multiply and adapt.

   随着技术的进步,工程学的分支领域大量增加且不断调整。

7. A mechatronics engineer unites the principles of mechanics, electronics, and computing to generate a simpler, more economical and reliable system.

   机电工程师将力学、电子和计算的原理结合起来,创造出一个更简单、更经济、更可靠的系统。

8. Electrical and electromechanical science and technology deals with the generation, distribution, switching, storage, and conversion of electrical energy to and from other energy forms.

   电气和机电科学技术处理的是电能的产生、分配、转换、存储,以及电能与其他能源的转换。

9. Most electronic devices use semiconductor components to perform electron control.

   大多数电子设备使用半导体元件来执行电子控制。

10. If you can get it into prototype, we can evaluate it.

    如果你能创建原型的话,我们可以对其进行评估。

11. We can use these tests as a simple prototype of how they might implement their coordinator.

    我们可以将这些测试作为实现协调程序的简单原型使用。

12. The company showed the prototype of the new model at the exhibition.

    公司在展览会上展示了这种新模型的原型。

13. Our next step is to gather feedback on the prototype and the ideas behind it.

    我们下一步是收集有关原型及其背后设想的反馈信息。

14. It is possible to make some rough estimates about the Analytical Engine's capability.

    对这台分析机的性能做粗略的评估是可能的。

15. What new concept in computing was introduced in the design of Babbage's Analytical Engine?

巴贝奇设计的分析机引入了哪些新的计算概念?

16. Babbage thought of the concept of software and hired the first programmer for his analytical engine.

巴贝奇首次提出了软件的概念,并为他的解析机聘请了一个程序员。

17. Press this button to start the engine.

按这个按钮启动发动机。

18. I waited in the car while idling the engine.

发动机空转时我在车里等待。

19. Each worker is, in a way, a cog in this huge, well-oiled production machine.

在某种意义上,每个工人都是这台巨大生产石油机器中的齿轮。

20. The wheel engages with the cog and turns it.

轮子与轮齿啮合并带动它转动。

21. The front cog has 40 teeth.

前齿轮有 40 个齿。

22. The two cog wheels engaged and the machine started.

那两个齿轮一啮合,机器就启动了。

23. A pocket calculator only works to eight decimal places.

袖珍计算器只能计算到小数点后 8 位数。

24. The figure is accurate to two decimal places.

这个数精确到小数点后两位。

25. There is a very simple technique for converting any decimal number to its binary equivalent.

有一个很简便的方法可以将任意十进制数换算为等值的二进制数。

26. The machine does the calculations in binary.

这台机器采用二进制进行计算。

27. The binary system of numbers is used in digital computers.

数字计算机都使用二进制。

28. In fact, there are many programs that contain no arithmetic whatsoever.

事实上,许多程序中根本不包含算术运算。

29. Computers of this generation are characterized by more and more transistors being contained on a silicon chip.

这一代计算机的特征是一个芯片上包含越来越多的晶体管。

30. Does the electronic computer based on microprocessors made of silicon chip come to an end?

以硅片微处理器为核心的电子计算机要走到尽头了吗?

31. A dynamo is used to generate electricity.

发电机用于发电。

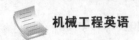

32. The dynamo can keep going with little fuel.

这台发电机用少量燃油就可以运转。

33. As soon as the dynamo slows down, the whole system will lag.

一旦这台发电机减速运转,整个系统都会慢下来。

34. I'm afraid the trouble is in the dynamo.

恐怕毛病出在发电机上。

35. The only accessory inside is a wall charger.

盒子内唯一的附件是一个壁式充电器。

36. The machine may lack a disc drive, but that doesn't mean you won't be able to get one — it will just come as an accessory that costs extra.

这台机器可能没有硬盘,但这并不意味你不能加一个——只不过需要为这个配件额外付钱。

37. A radio is an accessory to a car.

收音机是汽车的附件。

38. This company's business includes hardware trouble elimination, network connection, and accessory assembling, etc.

这家公司的业务包括硬件方面的故障排除、网络连接、散件组装等。

39. You can follow this procedure for any kind of resource: configuration file, audio file, graphics file, you name it.

你可以按照此步骤寻找任何类型的资源:配置文件、语音文件、图形文件,只要是你能说出来的。

40. This is a computer manufacturer which specializes in graphics.

这是一家专注于图形设计的计算机制造商。

41. Scientists currently debate the future implications of nanotechnology.

科学家们目前正在讨论纳米技术未来意味着什么。

42. Nanotechnology may be able to create many new materials and devices with a vast range of applications.

纳米技术也许能够创造出许多应用广泛的新材料和新设备。

43. Once you modify this file, upload it via RAM.

一旦修改这个文件,可通过随机存储器上传。

44. In this case, you will need at least 2GB of physical RAM to start both databases simultaneously.

在这种情况下,你需要至少 2GB 的物理 RAM,才能同时启动这两个数据库。

45. The control system has a good dynamic performance without current control loop.

尽管没有电流控制回路,但这个控制器仍具有较高的动态性能。

46. Every transistor has at least three electrodes.

每个晶体管至少有三个电极。

47. This means that even when the transistor is in the "off" state, a small amount of current

still flows through.

这意味着,即使晶体管处在"关闭"状态,少量的电流仍然能够流过。

48. The design is an important application of microcomputer feedback control system.

本设计是微机反馈控制系统的重要应用。

49. The application of planar linkage is very extensive in mechanical engineering.

平面连接在机械中的应用非常广泛。

50. The integrated design makes the work reliability of electric get improved greatly.

一体化设计使电气的工作可靠性得到大大提高。

Task Two Sample Dialogue

Directions: *In this dialogue, you are going to read several times the following sample dialogue about the relevant topic. Please pay special attention to five Cs (culture, context, coherence, cohesion and critique) in the dialogue and get ready for a smooth communication in the coming task.*

(*a lecture on mechatronics is just over*; *Mike and Steve went over to Prof. Smith.*)

Mike: Excuse me, Prof. Smith. Could you spare us several minutes?

Prof. Smith: Of course! So what's up?

Mike: We're a little confused about the difference between mechanics and mechatronics. How could we tell a machine is mechanical or mechatronical?

Prof. Smith: A good question, boy. The word "mechatronics" is a combination of mechanics and electronics; however, as technical systems have become more and more complex the definition has been broadened to include more technical areas, such as robotics, computer engineering, telecommunications engineering, systems engineering and control engineering. To put it simply, mechanics is a subfield of mechatronics.

Steve: Can I take it that the mechatronical products are more complicated than mechanical products?

Prof. Smith: It's often the case, but it depends. The mechanism of some mechatronical products is rather simple. Take the panel switch and mechanically-driven manipulator as examples. The former is mechatronical but simpler, while the latter is mechanical but more complex.

Mike: Oh I see.

Steve: Thank you, Prof. Smith. I have another question.

Prof. Smith: Come on, then.

Steve: Why do we name some machines numerical control machines?

Prof. Smith: Numerical control actually refers to the control of a machine tool or any other processing machines by using a series of mathematical information, or numerical data. It means the work of machines is controlled by a numerical control

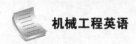

program.

Steve: I think I've got the point. Thank you very much, Prof. Smith.

Prof. Smith: You're welcome!

Task Three   Simulation and Reproduction

Directions: *The class will be divided into three major groups, each of which will be assigned a topic. In each group, some students may be the teacher, while others may be students. In the process of discussion, please observe the principles of cooperation, politeness and choice of words. One of the groups will be chosen to demonstrate the discussion to the class.*

1) electro-mechanics in our daily life

2) a funny story related to electromechanical product in my childhood

3) the importance of electromechanical integrated design

Task Four   Discussion and Debate

Directions: *The class will be divided into two groups. Please choose your stand in regard to the following controversy and support your opinions with scientific evidences. Please refer to the specialized terms and classical sentences in the previous parts of this unit.*

Technological development has brought great progress and convenience to our life. Advanced machines have taken a lot of weight off man's shoulders. However, some others hold the opinion that development of technology has brought environmental pollution and resources exhaustion and that people are working under greater pressure than before. What's your opinion? Please give your reasons.

V. After-class Exercises

1. *Match the English words in Column A with the Chinese meaning in Column B.*

A	B
1) actuator	A) 自动的;汽车的
2) mechanism	B) 电流
3) simulation	C) 十进制的;小数的
4) memory	D) 附件
5) automotive	E) 模仿;模拟
6) transistor	F) 发电机
7) generator	G) 结构,机械装置
8) decimal	H) 促动器
9) current	I) 晶体管

10) accessory   J）存储器,内存

2. *Fill in the following blanks with the words in the word bank. Change the form if it's necessary.*

binary	interact	transmission	battery	graphics
simulate	electrify	device	transistor	thermostat

1) BBC has just successfully demonstrated a new digital radio system _____.

2) Would you please tell me where I can get _____ for this brand of camera? It has run out of power.

3) A computer is a(n) _____ for processing information.

4) If we _____ the electronic device, the electric current will cause the electrons to eject quickly.

5) Today the report states, broadly, that the number of _____ on an integrated circuit will double roughly every two years.

6) Millions of people want new, simplified ways to _____ with a computer.

7) Is it reasonable to suggest that a(n) _____ has feelings? So that it can adjust the temperature automatically.

8) The scientist developed one model to _____ a full year of the globe's climate.

9) I am certain that they are referring to _____ digits and not decimal ones.

10) But computer _____ now enable us to shift the picture as the viewpoint shifts.

3. *Rewrite the follow sentences after the model.*

   Model: It was a machine. The machine was never built.

   →It was a machine that was never built.

1) The young lady is from 21st Talent Net. The young lady is interviewing Lin Shiying.

   _____

   _____

2) I'm looking for the watch. I bought the watch last week.

   _____

   _____

3) Do you know the girl? Our English teacher often talks with the girl.

   _____

   _____

4) Yesterday Li Ming visited the village. His family lived in the village ten years ago.

   _____

   _____

5) I was standing next to a pretty girl. Her name is Diana.

   _____

   _____

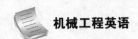

4. *Translate the following sentences into English.*

1）在19世纪30年代和40年代,简单的计算机被创造出来,开始了我们今天的计算机革命。

_____

_____

2）我真的很怀念那个时代,你可以去参加一个聚会,然后看到一台机械计算机展示在你面前。

_____

_____

3）雅卡尔(Jacquard)发明了提花织机,可以通过控制穿孔卡片来编织不可思议的图案。

_____

_____

4）这台机械打印机只能打印数字,因为它的设计者痴迷于数字。

_____

_____

5）他从来没有制造出自己设计的机器,他总是在做新计划。

_____

_____

5. *Please write an essay of about 120 words on the topic*：**Application of mechatronics in our life.** *Some specific examples will be highly appreciated and you have to watch out the spelling of some specialized terms you have learnt in this unit.*

_____

_____

_____

_____

_____

_____

_____

_____

VI. Additional Reading

Introduction to the Development History of General Electric Company

General Electric Company（GE）is an American multinational conglomerate（联合大公司,企业集团）incorporated in New York and headquartered in Boston. As of 2018, the company operates through the following segments: aviation (航空), healthcare, power, renewable energy, digital, additive manufacturing, venture capital and finance, lighting, transportation, and oil and gas.

In 2018, GE ranked among the Fortune 500 (财富 500 强) as the 18th-largest firm in the U. S. by gross revenue (收入总额). In 2011, GE ranked among the Fortune 20 as the 14th-most profitable company. As of 2012, the company was listed as the fourth-largest in the world among the Forbes Global 2000, further metrics being taken into account. Two employees of GE have been awarded the Nobel Prize: Irving Langmuir in 1932 and Ivar Giaever in 1973.

General Electric Building at 570 Lexington Avenue, NY

Formation

During 1889, Thomas Edison had business interests in many electricity-related companies including Edison Lamp Company, a lamp manufacturer in East Newark, New Jersey; Edison Machine Works, a manufacturer of dynamos and large electric motors in Schenectady, New York; Bergmann & Company, a manufacturer of electric lighting fixtures (照明灯具), sockets, and other electric lighting devices; and Edison Electric Light Company, the patent-holding company and the financial arm backed by J. P. Morgan and the Vanderbilt family for Edison's lighting experiments.

In 1889, Drexel, Morgan & Co., a company founded by J. P. Morgan and Anthony J. Drexel, financed Edison's research and helped merge those companies under one corporation to form Edison General Electric Company, which was incorporated in New York on April 24, 1889. The new company also acquired Sprague Electric Railway & Motor Company in the same year.

In 1880, Gerald Waldo Hart formed the American Electric Company of New Britain, Connecticut, which merged a few years later with Thomson-Houston Electric Company, led by Charles Coffin. In 1887, Hart left to become superintendent (主管) of the Edison Electric Company of Kansas City, Missouri. General Electric was formed through the 1892 merger of Edison General Electric Company of Schenectady, New York, and Thomson-Houston Electric Company of Lynn, Massachusetts, with the support of Drexel, Morgan & Co. Both plants continue to operate under the GE banner to this day. The company was incorporated in New York, with the Schenectady plant used as headquarters for many years thereafter. Around the same time, General Electric's Canadian counterpart, Canadian General Electric, was formed.

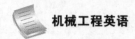

Public company

In 1896, General Electric was one of the original 12 companies listed on the newly formed Dow Jones Industrial Average(道琼斯工业平均指数,简称"道指"), where it remained a part of the index for 122 years, though not continuously.

General Electric in Schenectady, NY, aerial view, 1896

In 1911, General Electric absorbed the National Electric Lamp Association (NELA) into its lighting business. GE established its lighting division headquarters at Nela Park in East Cleveland, Ohio. The lighting division has since remained in the same location.

RCA and NBC

Owen D. Young, through GE, founded the Radio Corporation of America (RCA) in 1919, after purchasing the Marconi Wireless Telegraph Company of America. He aimed to expand international radio communications. GE used RCA as its retail arm for radio sales. In1926, RCA co-founded the National Broadcasting Company (NBC), which built two radio broadcasting networks. In 1930, General Electric was charged with antitrust violations(反垄断违法行为) and decided to divest itself of RCA.

Television

In 1927, Ernst Alexanderson of GE made the first demonstration of his television broadcasts at his home at 1132 Adams Rd, Schenectady, New York. On January 13, 1928, he made what was said to be the first broadcast to the public in the United States on GE's W2XAD: the pictures were picked up on 1.5 square inch (9.7 square centimeters) screens in the homes of four GE executives. The sound was broadcast on GE's WGY (AM).

Experimental television station W2XAD evolved into station WRGB which, along with WGY and WGFM (now WRVE), was owned and operated by General Electric until 1983.

Power generation

Led by Sanford Alexander Moss, GE moved into the new field of aircraft turbo superchargers (飞机涡轮增压器). GE introduced the first superchargers during World War I, and continued to develop them during the interwar period. Superchargers became indispensable in the years immediately prior to World War II. GE supplied 300,000 turbo superchargers for use in fighter and bomber engines. This work led the U. S. Army Air Corps(美国陆军航空兵) to select GE to develop the nation's first jet engine during the war. This experience, in turn, made GE a natural selection to develop the Whittle W. 1 jet engine that was demonstrated in the United States in 1941. GE ranked ninth among United States corporations in the value of wartime production contracts. Although their early work with Whittle's designs was later handed to Allison Engine Company, GE Aviation emerged as one of the world's largest engine manufacturers, by passing(绕过,避开) the British company Rolls-Royce plc.

Some consumers boycotted(联合抵制) GE light bulbs, refrigerators and other products in the 1980s and 1990s to protest GE's role in nuclear weapons production.

In 2002, GE acquired the wind power assets of Enron during its bankruptcy proceedings. Enron Wind was the only surviving U. S. manufacturer of large wind turbines (风力涡轮机) at the time, and GE increased engineering and supplies for the Wind Division and doubled the annual sales to $1.2 billion in 2003. It acquired ScanWind in 2009.

Computing

GE was one of the eight major computer companies of the 1960s along with IBM, Burroughs (宝来公司), NCR(国营收银机公司), Control Data Corporation(控制数据公司), Honeywell(霍尼韦尔公司), RCA(美国无线电公司) and UNIVAC(尤尼克公司). GE had a line of general purpose and special purpose computers, including the GE 200, GE 400, and GE 600 series general purpose computers, the GE 4010, GE 4020, and GE 4060 real-time process control computers, and the DATANET - 30 and Datanet 355 message switching computers. A Datanet 500 computer was designed, but never sold.

In 1962, GE started developing its GECOS(通用电气公司综合操作系统) (later renamed GCOS) operating system, originally for batch processing(成批处理), but later extended to timesharing and transaction processing. Versions of GCOS are still in use today. From 1964 to 1969, GE and Bell Laboratories (which soon dropped out) joined with MIT to develop the Multics Operating System (多操作系统) on the GE 645 mainframe computer. The project took longer than expected and was not a major commercial success, but it demonstrated concepts such as single level store(单层存储), dynamic linking(动态连接), hierarchical file system (分级文件系统), and ring-oriented security(环向安全). Active development of Multics continued until 1985.

GE got into computer manufacturing because in the 1950s they were the largest user of computers outside the United States federal government, aside from being the first business in

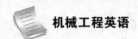

the world to own a computer. Its major appliance manufacturing plant "Appliance Park" was the first non-governmental site to host one. However, in 1970, GE sold its computer division to Honeywell, exiting the computer manufacturing industry, though it retained its timesharing operations for some years afterwards. GE was a major provider of computer time-sharing services, through General Electric Information Services (GEIS, now GXS), offering online computing services that included GEnie.

In 2000 when United Technologies Corp. planned to buy Honeywell, GE made a counter-offer that was approved by Honeywell. On July 3, 2001, the European Union issued a statement that "prohibits the proposed acquisition by General Electric Co. of Honeywell Inc. ". The reasons given were it "would create or strengthen dominant positions on several markets and that the remedies proposed by GE were insufficient to resolve the competition concerns resulting from the proposed acquisition of Honeywell. "

On June 27, 2014, GE partnered with collaborative design company Quirky to announce its connected LED bulb called Link. The Link bulb is designed to communicate with smart phones and tablets using a mobile app called Wink.

Major acquisitions and divestments

In 1986, GE reacquired RCA, primarily for the NBC television network (also parent of Telemundo Communications Group). The remainder was sold to various companies, including Bertelsmann (Bertelsmann acquired RCA Records) and Thomson SA, which traces its roots to Thomson-Houston, one of the original components of GE. Also in 1986, Kidder, Peabody & Co. , a U. S. -based securities firm(证券公司), was sold to GE and following heavy losses was sold to PaineWebber in 1994.

In 2004, GE bought 80% of Universal Pictures from Vivendi(威望迪环球影业). Vivendi bought 20% of NBC forming the company NBC Universal(美国国家广播环球公司). GE then owned 80% of NBC Universal and Vivendi owned 20%. By January 28, 2011 GE owned 49% and Comcast 51%. On March 19, 2013, Comcast(康卡斯特公司,美国最大的有线电视传输和宽带通信公司) bought GE's shares in NBCU(国家伤亡保险局) for $16.7 billion. In 2004, GE completed the spin-off(副产品) of most of its mortgage(抵押,债权) and life insurance assets(人寿保险资产) into an independent company, Genworth Financial, based in Richmond, Virginia.

Genpact(简柏特) formerly known as GE Capital International Services (GECIS) was established by GE in late 1997 as its captive India-based BPO. GE sold 60% stake in Genpact to General Atlantic and Oak Hill Capital Partners in 2005 and hived off(分群) Genpact into an independent business. GE is still a major client to Genpact today, for services in customer service, finance, information technology and analytics.

In May 2007, GE acquired Smiths Aerospace for $4.8 billion. Also in 2007, GE Oil & Gas acquired Vetco Gray for $1.9 billion, followed by the acquisition of Hydril Pressure & Control in 2008 for $1.1 billion.

On December 3, 2009, it was announced that NBC Universal would become a joint venture between GE and cable television operator Comcast. Comcast would hold a controlling interest in the company, while GE would retain a 49% stake and would buy out shares owned by Vivendi. Vivendi would sell its 20% stake in NBC Universal to GE for US $5. 8 billion. Vivendi would sell 7.66% of NBC Universal to GE for US $2 billion if the GE/Comcast deal was not completed by September 2010 and then sell the remaining 12.34% stake of NBC Universal to GE for US $3. 8 billion when the deal was completed or to the public via an IPO if the deal was not completed.

In April 2014, it was announced that GE was in talks to acquire the global power division of French engineering group Alstom for a figure of around $13 billion. A rival joint bid was submitted in June 2014 by Siemens and Mitsubishi Heavy Industries (MHI)(三菱重工) with Siemens seeking to acquire Alstom's gas turbine(涡轮机;汽轮机) business for 3.9 billion, and MHI proposing a joint venture in steam turbines, plus a 3.1 billion cash investment. In June 2014 a formal offer from GE worth $17 billion was agreed by the Alstom board. Part of the transaction involved the French government taking a 20% stake in Alstom to help secure France's energy and transport interests and French jobs. A rival offer from Siemens-Mitsubishi Heavy Industries was rejected. The acquisition was expected to be completed in 2015. In October 2014, GE announced it was considering the sale of its Polish banking business Bank BPH.

Later in 2014, General Electric announced plans to open its global operations center in Cincinnati, Ohio. The Global Operations Center opened in October 2016 as home to GE's multifunctional shared services organization. It supports the company's finance/accounting, human resources, information technology, supply chain, legal and commercial operations, and is one of GE's four multifunctional shared services centers worldwide in Pudong, China; Budapest, Hungary; and Monterrey, Mexico.

GE Global Operation Center in Downtown Cincinnati, Ohio

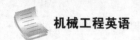

In April 2015, GE announced its intention to sell off its property portfolio(房地产投资组合), worth $26.5 billion, to Wells Fargo and the Blackstone Group. It was announced in April 2015 that GE would sell most of its finance unit and return around $90 billion to shareholders as the firm looked to trim down(削减) on its holdings and rid itself of its image of a "hybrid" company, working in both banking and manufacturing. In August 2015, GE Capital agreed to sell its Healthcare Financial Services business to Capital One for US $9 billion. The transaction involved US $8.5 billion of loans made to a wide array of sectors including senior housing, hospitals, medical offices, outpatient services(门诊服务), pharmaceuticals(医药品,药物) and medical devices. Also in August 2015, GE Capital agreed to sell GE Capital Bank's on-line deposit platform to Goldman Sachs. Terms of the transaction were not disclosed, but the sale included US $8 billion of on-line deposits and another US $8 billion of brokered certificates (代理证书) of deposit. The sale was part of GE's strategic plan to exit the U. S. banking sector and to free itself from tightening banking regulations. GE also aimed to shed its status as a "systematically important financial institution".

In September 2015, GE Capital agreed to sell its transportation-finance unit to Canada's Bank of Montreal. The unit sold had US $8.7 billion (CA $11.5 billion) of assets, 600 employees and 15 offices in the U. S. and Canada. Exact terms of the sale were not disclosed, but the final price would be based on the value of the assets at closing, plus a premium according to the parties. In October 2015, activist investor Nelson Peltz's fund Trian bought a $2.5 billion stake in the company.

In January 2016, Haier Group acquired GE's appliance division for $5.4 billion. In October 2016, GE Renewable Energy agreed to pay 1.5 billion to Doughty Hanson & Co for LM Wind Power during 2017.

At the end of October 2016, it was announced that GE was under negotiations for a deal valued at about $30 billion to combine GE Oil and Gas with Baker Hughes. The transaction would create a publicly-traded entity controlled by GE. It was announced that GE Oil and Gas would sell off its water treatment business as part of its divestment agreement with Baker Hughes. The deal was cleared by the EU in May 2017, and by the DOJ in June 2017. The merger agreement was approved by shareholders at the end of June 2017. On July 3, 2017, the transaction was completed and Baker Hughes became a GE company.

In April 2017, GE announced the name of their $200 million corporate headquarters would be "GE Innovation Point". The groundbreaking ceremony for the 2.5 – acre, 800 – person campus was held on May 8, 2017, and the completion date is expected to be sometime in mid−2019.

(*If you want to find more information about this corporation, please log on https://en. wikipedia. org/wiki/General_Electric.*)

1. *Read the passage quickly by using the skills of skimming and scanning, and choose the best*

answer to each of the following questions.

1) GE ranked among the Fortune 20 as the 14th-most profitable company in _____.
   A. 2000   B. 2011
   C. 2012   D. 2018

2) Which of the following was NOT included in the electricity-related companies in which Thomas Edison had business interests? _____
   A. Edison Lamp Company   B. Edison Machine Works
   C. Bergmann & Company   D. Drexel, Morgan & Co

3) General Electric was formed through _____.
   A. the formation of Edison General Electric Company in 1889
   B. Edison General Electric Company's acquisition of Sprague Electric Railway& Motor Company in 1889
   C. the merger of American Electric Company and Thomson-Houston Electric Company in 1880s
   D. the merger of Edison General Electric Company and Thomson-Houston Electric Company in 1892

4) Originally only _____ companies were listed on the Dow Jones Industrial Average.
   A. 11   B. 12
   C. 13   D. 14

5) In 1919, Owen D. Young, founded the Radio Corporation of America(RCA), aiming to _____.
   A. expand international radio communications
   B. become the retail arm of GE for radio sales
   C. build more profitable radio broadcasting networks
   D. attract more foreign investment

6) _____ directly led the U. S. Army Air Corps to select GE to develop the nation's first jet engine during the World War II.
   A. GE's entry into the new field of aircraft turbo superchargers
   B. GE supplying turbo superchargers for use in fighters and bomber engines
   C. GE's introduction of the first superchargers during World War I
   D. GE's efforts in developing the superchargers during the interwar period

7) In the computing field, GE once worked jointly with Bell Laboratories and MIT to develop an operating system named _____.
   A. GECOS   B. GCOS
   C. Multics   D. GE 645

8) GE got into computer manufacturing because _____.
   A. in the 1950 they were the largest user of computers besides the U. S. federal government
   B. they wanted to be the first business in the world to own a computer

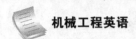

C. they expected a major commercial success from its computer division

D. it was profitable

9) In which acquisition GE encountered a rival joint bid from Siemens-Mitsubishi Heavy Industries? _____

A. the reacquisition of RCA

B. the acquisition of Universal Pictures from Vivendi

C. the acquisition of Smiths Aerospace

D. the acquisition of the global power division of French engineering group Alstom

10) It was announced in 2015 that GE would sell most of its finance unit and return around $90 billion to shareholders because _____.

A. it tried to whittle down its holdings

B. it wanted to get rid of the image of a hybrid company

C. the finance division got involved in debts

D. both A and B

2. *In this part*, *the students are required to make an oral presentation on either of the following topics.*

1) Introduce a GE's electromechanical product you are familiar with.

2) Simulate a situational dialogue for an interview to GE.

习题答案

Unit Five Industrial Design

I. Pre-class Activity

Directions: *Please read the general introduction about* **Dieter Rams** *and tell something more about the great scientist to your classmates.*

Dieter Rams (1932—   ) is a German industrial designer. His unobtrusive approach and belief in "less but better" design generated a timeless quality in his products and have influenced the design of many products, which also secured Rams worldwide recognition and appreciation.

Rams and his staff designed many memorable products for Braun including the famous SK – 4 record player and the high-quality 'D'-series (D45, D46) of 35mm film slide projectors. He is also known for designing a furniture collection for Vitsoe(英国家具公司名) in the 1960s including the 606 universal shelving system and 620 chair programs. By producing electronic gadgets that were remarkable in their austere aesthetic and user friendliness, Rams made Braun a household name in the 1950s.

Rams introduced the idea of sustainable development and of obsolescence being a crime in design in the 1970s. Accordingly, he asked himself the question: "Is my design good design?" The answer he formed became the basis for his celebrated ten principles. According to them, "good design": ①is innovative; ②makes a product useful; ③is aesthetic; ④makes a product understandable; ⑤is unobtrusive; ⑥is honest; ⑦is long-lasting; ⑧is thorough down to the last detail; ⑨is environmentally friendly; ⑩is as little design as possible.

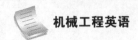

II. Specialized Terms

Directions: *Please remember the following specialized terms before the class so that you will be able to cope with the coming tasks better.*

adjustable adj. 可调整的

aesthetic adj. 美(学)的;审美的;具有审美趣味的 n. 美感;审美观

aesthetics n. 美学;美术理论

applicable adj. 适当的;可应用的

artifact n. 人工制品,手工艺品,加工品

artist n. 艺术家;画家;能手

blueprint n. 蓝图,设计图

brand development 品牌发展

CAD model 计算机辅助设计模型

chief engineer 主任工程师;总工程师

clay model 泥塑模型;土模

computer-aided industrial design 计算机辅助工业设计

concept design 概念设计

conceptual design 概念设计,方案设计

conceptualization n. 化为概念,概念化

conveyer system 流水作业

craft n. 手艺;飞行器;船 vt. 手工制作;精巧地制作

craft design 工艺美术设计

craftsmanship n. 技术,技艺

creative director 创意总监

customer need 顾客需要

customer requirement 客户需求

decorative adj. 装饰的;装潢用的

decorator n. 室内装饰师,油漆匠 adj. 适于室内装饰的

deployment n. 部署;调度

design director 设计总监

design review 设计评论

detailed design 详细设计

development cycle 开发周期

digital model 数字模型

diversification n. 变化,多样化;多种经营

durable adj. 耐用的;持久的 n. 耐用品,耐久品

dynamic design 动态设计

eco-design 生态(化)设计;环境化设计

eco-friendly 生态友好型,自然环保的

electronic gadget 电子装置

embodiment design 实体化设计

engraver n. 雕刻师,雕工

ergonomics n. 工效学;人类工程学

external surface treatment 外表面处理

fastener n. 紧固件

feasibility assessment 可行性估计

foam model 泡沫模型

functional analysis 功能分析

functional design 功能设计

functional differentiation 功能分化

functionality n. 功能;功能性;设计目的;实用

generalized industrial design 广义工业设计

human-centered design 人性化设计

humanize v. 赋予人性;变为有人性

IDSA *abbr.* Industrial Designers Society of America 美国工业设计家协会

industrial designer 工业设计师,产品设计师

innovative adj. 革新的;创新的

intellectual property 知识产权

interchangeable parts 可互换性零件

interior design 内部装饰业;室内设计

life cycle 生活(生命)周期

lightweight adj. 轻量的

manual adj. 用手的;手制的,手工的 n. 手册;指南

manufacturability n.可制造性

mass production 大规模生产

material requirement 材料要求

metalwork n. 金属制品,金属制造

minimalism n. 极简派艺术

narrow industrial design 狭义工业设计

operating parameter 操作参数,运行参数

originality n. 独创性,创造性;匠心;独到之处;新颖

oscillation n. 振动;波动

packaging design 包装设计

parameter n. 参数;系数;参量

patent for invention 发明专利

performance n. 性能;绩效

pragmatic adj. 实际的;实用主义的

preliminary design 初步设计

product design 产品设计

product development 产品开发

product image 产品形象

product testing 产品测试

production planning 生产规划;生产制造

计划

production process 生产流程

prototype testing 样机试验;原型试验

qualification testing 合格性试验

reframe vt. 再构造,再组织

reverse engineering 逆向工程

sensory design 感官设计

service design 服务设计

set design 成套设计

solid model 实体模型

standardization n. 标准化

sustainability n. 持续性;永续性;能维持性

system engineering 系统工程

test engineer 测试工程师

three-dimension design 三维设计

titanium n. [化]钛

titanium alloy 钛合金

unit design 组合式设计

usability n. 合用,可用;可用性

user-centered design 以用户为中心的设计

user-friendly adj. 用户界面友好的,用户容易掌握使用的

visual design 视觉设计

well-conceived 构思良好的

III. Watching and Listening

Task One   Reconsider the Way We Sit（I）

New Word

aeronautics n. 航空学,飞行术

self-righting adj. 自动调节的,自动扶正的

fantasy n. 幻想;空想的产物 vt. & vi. 想象;幻想

fuselage n. <空>(飞机的)机身;火箭的外壳

aeronautical adj. 航空的

wile vt. 消遣;消磨

dole out 少量地发放(食物、救济金等)

interior n. 内部   adj. 内部的

tractor n. 牵引器;拖拉机

duplicate v. 复制

视频链接及文本

conceive vt. & vi. 构思;想象,设想　　　　engineer vt. 设计,策划

Exercises

1. *Watch the video for the first time and choose the best answer to each of the following questions.*

1) If you wanted the model airplane to fly, you had to learn the following EXCEPT _____.

　A. the discipline of flying

　B. knowledge about aeronautics

　C. what made an airplane stay in the air

　D. the approach of drawing the fantasy shapes of airplanes

2) The act of drawing airplanes led the speaker to _____.

　A. build airplanes

　B. build model airplanes

　C. register for an aeronautical engineering course

　D. fall in love with airplanes

3) The speaker went into the art department because _____.

　A. he liked art

　B. he thought his drawing would be appreciated there

　C. his romance was changed

　D. he fell in love with a girl from art department

4) For the 25 years after his graduation, the speaker worked as _____.

　A. an artist　　　　　　　　　B. a designer

　C. an architect　　　　　　　 D. a painter

5) In order to get back the romance, the speaker turned to _____.

　A. design office chairs

　B. furniture design

　C. design chairs for tractors and submarines

　D. interior design for airplanes

2. *Watch the video again and decide whether the following statements are true or false.*

1) In the days when the speaker was a young man, a model airplane had to be self-righting. (　)

2) The speaker found the aeronautical engineering course very interesting. (　)

3) The speaker didn't get the slightest satisfaction from the objects he designed for 25 years. (　)

4) The speaker's romance began with airplanes but later he found them sort of unromantic. (　)

5) According to the speaker, his design work on the airplanes and that on the chair share something in common. (　)

3. *Watch the video for the third time and fill in the following blanks.*

So after 25 years I began to feel _____ as though I was running dry. And I quit. And I started up a very small _____—went from 40 people to one—in an effort to _____ my innocence. I wanted to get back where the romance was. I couldn't choose airplanes because they had gotten sort of _____ at that point. Even though I'd done a lot of airplane work on the _____. So I chose _____. And I chose chairs because I knew something about them. I'd designed a lot of chairs, over the years for tractors and trucks and submarines. All kinds of things. But not _____. So I started doing that. And I found that there were ways to _____ the same approach that I used to use on the airplanes. Only this time instead of the product being shaped by the wind, it was shaped by the _____. So the _____ was, as in the airplane you learn a lot about how to deal with the air, for a chair you have to learn a lot about how to deal with the body.

4. *Share your opinions with your partners on the following topics for discussion.*

1) What is the stimulus for you to choose your major? And what have you obtained from the courses you take? Select one subject to talk about.

2) If you were supposed to design something for your company, what would you do to complete the task?

Task Two   Reconsider the Way We Sit (II)

视频链接及文本

New Words

preconceived adj. 预想的,先入为主的

notion n. 概念,观念;意见

style v. 设计;为……造型

humanly adv. 在人力所及的范围,依靠人力

mechanistically adj.机械地

fuss with 过分讲究

recline vi. 斜倚,倚靠 vt. 使躺下;使斜倚

joint n. 关节;接合处

trade-off n. 交易;权衡

accommodate vt. 容纳;使适应

bulk n. (大)体积;大块,大量

compromise n. /vi 妥协;折中 racetrack n. 跑道

lever n. 杠杆;操作杆;工具 vt. 用杠杆撬动

wristwatch n. 手表,腕表

armrest n. (座位的)靠手,扶手

in parallel 并行的,平行的

incrementally adv. 逐渐地

arch n. 弓形;拱门;拱形物

tailbone n. 尾骨

spine n. 脊柱;脊椎

breathability n. 透气性

headrest n. 靠头之物

vertical adj. 垂直的,竖立的

knob n. 球形把手

Exercises

1. *Watch the video for the first time and choose the best answer to the following questions.*

1) How long did it take the speaker to design this chair? _____

    A. a couple of years   B. two or three years

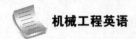 机械工程英语

C. four or five years   D. eight or nine years

2) According to the speaker, most good chair recline because _____.

A. they are beneficial for better breathing and better blood flow

B. they are easy to control

C. they can automatically balance your weight

D. they can automatically adjust height

3) Which is one of the drawbacks to the chair? _____

A. It's not suitable for very light people.

B. It's not suitable for people with bulky upper body.

C. Some people may fall off the chair.

D. People don't know how to adjust it.

4) In the speaker's opinion, a 20-page manual about how to use a chair is _____.

A. necessary   B. totally unnecessary

C. very helpful   D. amazing

5) About the design of the headrest of the chair, which of the following is true? _____

A. It's still a challenge to be solved.

B. It is in the same position no matter you're upright or reclined.

C. You need to turn a knob and adjust it.

D. The headrest will automatically hold your head in a vertical position when you recline.

2. *Watch the video again and decide whether the following statements are true or false.*

1) The speaker started designing this chair by drawing its sketches. (   )

2) Each individual armrest needs adjusting so that you can sit in the chair to work comfortably. (   )

3) The time the speaker spent in designing the armrest was much longer than in designing the headrest. (   )

4) Relieving your arms with armrests in the chair can take off 20% of the load from your upper body. (   )

5) The chair actually was human-centered and mechanistically automatic. (   )

3. *Watch the video for the third time and fill in the following blanks.*

When I started this chair it was not a preconceived _____. Design nowadays, if you mean it, you don't start with _____ sketches. I started with a lot of _____ ideas roughly eight or nine years ago. And these ideas had something to do with what I knew happened with people _____. At workplace, people who worked, and used task seating, a great many of them sitting in front of _____ all day long. And I felt, the one thing they don't need, is a chair that _____ with their main reason for sitting there. So I took the _____ that the chair should do as much for them as _____ possible or as mechanistically possible so that they didn't have to _____ with it. So my idea was that instead of sitting

down and reaching for a lot of controls, that you would sit on the chair, and it would
_____ balance your weight against the force required to recline.

4. *Share your opinions with your partners on the following topics for discussion.*

   1) If you were to design a piece of furniture, what would you take into consideration? And how would you resolve it?

   2) For an industrial designer, some special qualities or skills or abilities to design are necessary. What are they?

IV. Talking

Task One Classical Sentences

Directions: *In this section, some popular sentences are supplied for you to read and to memorize. Then, you are required to simulate and produce your own sentences with reference to the structure.*

General Sentences

1. I'm going shopping because I need to buy some clothes.
   我想去逛街,因为我需要买一些衣服。

2. Yesterday was such a beautiful day and we decided to go for a drive.
   昨天天气很好,我们开车出去玩了一趟。

3. What are you going to wear today?
   你今天打算穿什么?

4. I'm going to wear my blue suit. Is that all right?
   我要穿蓝色西装。怎么样?

5. I have some shirts to send to the laundry.
   我有一些衬衫要送到洗衣房去。

6. You ought to have that coat cleaned.
   你应该把那件外套洗一下。

7. I've got to get this shirt washed and ironed.
   我得把这件衬衫洗一洗,烫一烫。

8. All my suits are dirty. I don't have anything to wear.
   我所有的西装都脏了,我没有穿的了。

9. You'd better wear a light jacket. It's chilly today.
   你最好穿件薄夹克,今天很冷。

10. This dress doesn't fit me anymore.
    这件衣服我穿已经不合身了。

11. These shoes are worn-out. They've lasted a long time.

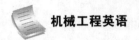

这些鞋我穿好久了,已经磨破了。

12. Why don't you get dressed now? Put on your work clothes.
    你干嘛还不换装? 穿上你的工作服呀。

13. My brother came in, changed his clothes, and went out again.
    我哥哥进来了,换好衣服后又出门了。

14. I didn't notice you were wearing your new hat.
    我都没注意到你戴了新帽子。

15. If you want a towel, look in the linen closet.
    如果你想要毛巾,到亚麻壁橱里找。

16. My brother wants to learn how to dance.
    我弟弟想学跳舞。

17. Which would you rather do—go dancing or go to cinema?
    你想干什么? 去跳舞还是去看电影?

18. I'd like to make an appointment to see Mr. Cooper.
    我想约时间去看望库珀先生。

19. Would you like to arrange for a personal interview?
    你想安排一场个人采访吗?

20. Your appointment will be next Thursday at 10 o'clock.
    你的会面安排在下周四 10 点。

21. I can come any day except Thursday.
    除了星期四,我都能来。

22. He wants to change his appointment from Monday to Wednesday.
    他想将会面从周一改到周三。

23. She failed to call the office to cancel her appointment.
    她没能打电话让办公室取消她的预约。

24. I'm going to call the employment agency for a job.
    我要打电话给职业介绍所找一份工作。

25. Please fill in this application form.
    请填写这张申请表。

26. Are you looking for a permanent/temporary position?
    你想应聘一个长期/临时职位吗?

27. I'm going to call a plumber to come this afternoon.
    我打算今天下午叫个管道工来。

28. I couldn't keep the appointment because I was sick.
    我病了,不能按时赴约。

29. Please call before you come, otherwise we might not be home.
    请在来之前打个电话,不然我们有可能不在家。

30. Will you please lock the door when you leave?

你离开时把门锁上,好吗?

31. I went to see my doctor for a check-up yesterday.
我昨天去医生那儿做了检查。

32. The doctor discovered that I'm a little overweight.
医生发现我有些超重。

33. He gave me a chest X-ray and took my blood pressure.
他让我做了个 X 光胸透,又给我量了量血压。

34. He told me to take these pills every four hours.
他叮嘱我每四小时吃一次药。

35. Do you think the patient can be cured?
你觉得这个病人能治愈吗?

36. —What did the doctor say?
—The doctor advised me to get plenty of exercise.
——医生说什么?
——医生建议我多做些运动。

37. If I want to be healthy, I have to stop smoking.
如果我想健康的话,我得戒烟。

38. It's just a mosquito bite. There's nothing to worry about.
只是被蚊子叮了下,不必担心。

39. —How are you feeling today?
—Couldn't be better.
——今天感觉怎样?
——非常好。

40. I don't feel very well this morning.
今天早上我感到不舒服。

41. I was sick yesterday, but I'm better today.
昨天我病了,不过今天好些了。

42. My fever is gone, but I still have a cough.
我已经不发烧了,不过还在咳嗽。

43. —Which of your arms is sore?
—My right arm hurts. It hurts right here.
——你的哪个胳膊疼?
——我的右胳膊受伤了,就在这儿。

44. —What's the matter/wrong/the trouble with you?
—I've got a pain in my back.
——你哪里不舒服?
——我背疼。

45. —How did you break your leg?

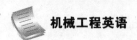

机械工程英语

—I slipped on the stairs and fell down. I broke my leg.

——你的脚怎么伤的？

——我在楼梯上摔了一跤,脚就受伤了。

46. Your right hand is swollen. Does it hurt?

你的右手肿了,疼吗？

47. It's bleeding. You'd better go to see a doctor about that cut.

你的伤口在流血,最好去医生那儿看看。

48. Why do you dislike the medicine so much?

为什么你这么不喜欢药？

49. I didn't like the taste of the medicine, but I took it anyway.

我不喜欢药的味道,但我还是吃了。

50. I hope you'll be well soon.

希望你快些好。

Specialized Sentences

1. He materialized his ideas by building a model.

他制作了一个模型使自己的想法具体化了。

2. This plane model miniatures the airplane.

这个飞机模型是飞机的缩影。

3. This model is not accurate.

该模型是不精确的。

4. You now have an empty concrete model.

你现在有了一个空的实物模型。

5. He devoted every spare moment to aeronautics.

他把他所有空余的时间都用在航空学上。

6. Some oscillation of the fuselage had been noticed on early flights.

在之前的几次飞行中就已经察觉到机身有些摇晃。

7. The wing section is connected to fuselage with titanium fasteners.

机翼部分用钛合金螺栓与机身连在一起。

8. The advent of aircraft brought with it aeronautical engineering.

宇宙飞船的问世使得航天工程出现。

9. Products are chosen for their aesthetic appeal as well as their durability and quality.

产品因其审美吸引力以及耐用性和质量好而被选中。

10. The interior is planned with a precision.

内部设计一丝不苟。

11. No one could conceive how such a machine could be constructed.

谁也想不通这样一台机器是怎么造出来的。

12. This spaceship was engineered by Bert Rutan.

94

这艘飞船由伯特·鲁坦设计建造。

13. From now on you should report to the chief engineer on all matters.

今后你的一切工作都要向总工程师报告。

14. As the test engineer, you can request the execution of an automation project on one or more lab resources.

作为测试工程师,你可以请求对一个或者更多的实验资源执行自动化项目。

15. In this test, the test engineer installs and sets up the software.

在此测试中,测试工程师安装和设置软件。

16. Pull the lever towards you to adjust the speed.

把操纵杆向你身体一侧拉动以调节速度。

17. Push hard and the lever will go down.

用力推就能把控制杆按下去。

18. The worst part of the set-up is the poor instruction manual.

安装时最糟糕的是操作指南讲述不够清楚。

19. In some cases, all of the steps in a process are manual.

在一些情况下,流程中的所有步骤都是手工的。

20. The temperature control knob doesn't work very well.

调温器的旋钮坏了。

21. Turn the knob to start the machine.

转动旋钮,启动机器。

22. Industrial design takes place in advance of the physical act of making a product.

工业设计发生在制造产品的物理行为之前。

23. Seat height, thigh support and backrest angle are all individually adjustable.

座椅高度、大腿支撑和靠背角度可进行个性化调整。

24. This blueprint is used to guide the phases of analysis, design, coding, deployment, maintenance, update, and so on.

设计图被用来指导分析、设计、编码、部署、维护、更新等阶段的工作。

25. All the machine parts on a blueprint must answer each other.

设计图上所有的机器部件都应互相配合。

26. The precision analysis of CAD Model is one of the key technologies that should be solved urgently in reverse engineering.

CAD 模型精度分析是逆向工程中亟待完善的关键技术。

27. I believe the device now actually looks better than the original concept design we published last summer.

我相信,现在的这款设备看起来比我们去年夏天发布的最初的概念设计更为出色。

28. The traditional mold design craft is too complex, and the development cycle is too long.

传统的模具设计工艺复杂,开发周期长。

29. Our products are favored by clients at home and abroad with the exquisite craft, novel

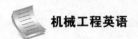

design and reliable quality.

我们的产品以工艺精湛、造型新颖、质量可靠赢得了中外客商的青睐。

30. They put forward the request of the product craft design.

他们提出工艺产品设计要求。

31. Rotor dynamic design is one of the most important aspects of aero-engine design.

转子动力学设计是航空发动机设计的一个重要环节。

32. This research mainly studies the analysis the theory and method of household electric appliance eco-design from designer's angle.

本研究主要是从设计师的角度对家用电器生态设计的理论和方法加以分析。

33. We now have extensive knowledge of overcoming the problems associated with eco-design.

我们现在拥有丰富的知识去克服与生态设计相关的问题。

34. Eco-design is a new design method that emerges in industry circles against the background of the growing gravity of environmental and resource problems.

生态设计是近年来在资源与环境问题的压力下,产业界正在兴起的一种设计方法。

35. In this process, our engineering design capabilities have been considerably increased.

在此过程中,我们的工程设计能力得到很大提高。

36. Our company has strong engineering design capabilities.

我公司有雄厚的工程设计能力。

37. They're designed with special attention to ergonomics, making typing easier.

它们的设计特别注重人体工程学,可以让打字更轻松。

38. Light touch accelerator pedal eliminates operator fatigue.

轻型的加速踏板减轻了操作者的疲劳。

39. Its technical feasibility needs further assessment and experiment.

其技术可行性有待进一步评估和实验。

40. The virtual model is cast into a new foam model, using the dimensions of the CAD model.

用 CAD 模型的尺寸,将虚拟模型转换为新的泡沫模型。

41. Functional analysis is an effective method in engineering to obtain a principle scheme.

功能分析是工程设计中探求原理方案的一种有效方法。

42. It is significantly more compact than any comparable one, with no loss in functionality.

它比任何同类产品都要紧凑得多,并且不会损失功能。

43. The added functionality means added complexity.

增加功能也意味着增加复杂性。

44. Try experimenting with combinations of these technologies to get the functionality that you want.

设法将这些技术结合在一起以得到你想要的功能。

45. Each new product would have a relatively long life cycle.

每一种新产品都会有一个相对较长的使用周期。

46. It might be launched into mass production in 2025.

它可能在 2025 年投入大批量生产。

47. The program plans to set new standards of excellence for product development and design.

本项目计划在产品开发和设计领域设定新的卓越标准。

48. Perhaps you dislike the visual design, or perhaps you want to augment some aspect of the functionality.

也许你不喜欢它的视觉设计,也许你想增强它某一方面的功能。

49. Based on this user feedback, designers tried to develop a different product to meet the market demands.

根据用户反馈,设计者们试图开发一种新产品来满足市场需求。

50. All these programs were in the phase of prototype testing.

所有这些程序均处于原型测试阶段。

Task Two　Sample Dialogue

Directions：*In this dialogue, you are going to read several times the following sample dialogue about the relevant topic. Please pay special attention to five Cs (culture, context, coherence, cohesion and critique) in the dialogue and get ready for a smooth communication in the coming task.*

A：So, Raphael, what did you study at university?

B：My main specialty is industrial design. Basically it is to design some products that are to be manufactured through techniques of mass production.

A：Sounds great! What courses did you have at your university?

B：I studied industrial art, modeling design, mechanics, CAD, material science, ergonomics and so on, and these courses qualified me to be an industrial designer. I am doing light and interior design for an auto manufacturer. That's my job!

A：I can imagine it must be creative and challenging. Do you like it?

B：Yes, I like my job. I'm really interested in interior design, especially furniture design for automobiles. That's the reason I come to Germany. Because they have some really interesting furniture designers here and they don't have the same strict architecture rules that we have in my country, like every time you want to build something new, or just make some arrangements for the auto interior, anything you need, once the project is done and everything is prepared, you need to show it to an architect or engineer who specializes in these very rules. He will tell you if you can or not actually build it or do it. He needs to consider whether your design corresponds to the rhythm of the existing objects, you know, the lines, the structure, and things like that.

A：It's really difficult. Things are changing very fast, and automobiles are no exception. So you need to adjust your design again and again to meet the architect or the engineer's requirements.

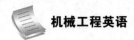

B: You said it. But here in Germany, there is a lot of liberty for interior designers. It's very free, they can do some really interesting things.

A: I bet every designer is eager to put thumb print on something he works on, so as to produce a moment, feeling, something like that, right?

B: Whatever it is, it is not important. What's important is, what you are creating has an effect.

A: I've got what you meant. Well, best of luck!

Task Three　Simulation and Reproduction

Directions: *The class will be divided into three major groups, each of which will be assigned a topic. In each group, some students may be the teacher, while others may be students. In the process of discussion, please observe the principles of cooperation, politeness and choice of words. One of the groups will be chosen to demonstrate the discussion to the class.*

1) the importance of industrial design

2) a funny story related to an industrial designer you admire

3) basic principles in industrial design

Task Four　Discussion and Debate

Directions: *The class will be divided into two groups. Please choose your stand in regard to the following controversy and support your opinions with scientific evidences. Please refer to the specialized terms and classical sentences in the previous parts of this unit.*

Color and shape are the two factors that industrial designers take into consideration when they are designing some new products. Some designers attach great importance to the functionality and practicability of the product so they emphasize the shape design. However, some other designers hold the opinion that good use of color can greatly enhance the visual appeal of the product and that color does a better job than shape in deepening the masses' impression of the product. What do you think? Please present more detailed information or examples to support your argument.

V. After-class Exercises

1. *Match the English words or phrases in Column A with the Chinese meaning in Column B.*

A	B
1) aesthetic	A) 性能;绩效
2) functionality	B) 视觉设计
3) performance	C) 工效学;人类工程学

4) visual design   D) 功能性;设计目的
5) dynamic design   E) 航空学,飞行术
6) craft design   F) 大规模生产,批量生产
7) well-conceived   G) 动态设计
8) mass production   H) 美(学)的;审美的
9) ergonomics   I) 工艺美术设计
10) aeronautics   J) 构思良好的

2. *Fill in the following blanks with the words in the word bank. Change the form if it's necessary.*

manual	intellectual	patent	conceptual	pragmatic
interior	manufacturability	originality	craft	blueprint

1) They started this design with some rough ideas, and they went from _____ design to working prototype in a year.

2) Copyright and trade secrets have historically been the primary protection mechanisms for software _____ property.

3) The plan is a(n) _____ for jobs and housing growth used by local authorities.

4) We must promote _____ and encourage innovation.

5) A(n) _____ is a book which tells you how to do something or how a piece of machinery works.

6) On Monday, India's Supreme Court rejected the firm's appeal to get _____ protection for the cancer drug.

7) Cost and _____ are of paramount importance to any manufacturer.

8) Being the _____ CEO that she is, Sankar got down to business—and created a plan.

9) By going digital, the _____ was quickly becoming democratized (普及).

10) Stylistically furniture is related to architecture and _____ design.

3. *Find a word from Part II Specialized Terms which means*:

1) inventing or contriving an idea or explanation and formulating it mentally

2) intended to look pretty or attractive, serving an esthetic rather than a useful purpose

3) the property of being able to continue or be maintained for a period of time

4) capable of being applied; having relevance

5) to support or enclose (a picture, photograph, etc.) in a new or different frame

6) being or producing something like nothing done or experienced or created before

7) the ability to think and act independently; the quality of being new, not derived from something else

8) existing for a long time; serviceable for a long time

4. *Translate the following sentences into English.*

1) 我不得不放弃异想天开的画法,把它转换成技术图纸。

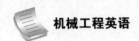

2）当我有机会在学校选择一门课程时,我决定报名学习航空工程。

3）他觉得自己似乎快才思枯竭(run dry)了,于是开始尝试新的东西,试图重新发现自己的价值。

4）我的想法是,你坐在椅子上,它会自动平衡你的体重和倾斜所需要的力量。

5）经过多次试验失败后,我们想出了一个非常简单的办法,只需要移动一只胳膊就能把扶手调到我们想要的位置。

5. *Please write an essay of about* 120 *words on the topic：* **The importance of industrial design.** *Some specific examples will be highly appreciated and you have to watch out the spelling of some specialized terms you have learnt in this unit.*

VI. Additional Reading

The World's Top Design Companies

1. IDEO

IDEO is an international design and consulting firm founded in Palo Alto, California, in

1991. The company has locations in Cambridge (Massachusetts), Chicago, London, Munich, New York City, Palo Alto, San Francisco, Shanghai and Tokyo. The company uses the design thinking methodology to design products, services, environments, and digital experiences. Additionally, the company has become increasingly involved in management consulting and organizational design. The firm employs over 700 people in a number of disciplines including: Behavioral Science, Branding, Business Design, Communication Design, Design Research, Digital Design, Education, Electrical Engineering, Environments Design, Food Science, Healthcare Services, Industrial Design, Interaction Design, Mechanical Engineering, Organizational Design, and Software Engineering.

History

IDEO was formed in 1991 by a merger of David Kelley Design (founded by Stanford University professor David Kelley), London-based Moggridge Associates and San Francisco's ID Two (both founded by British-born Bill Moggridge), and Matrix Product Design (founded by Mike Nuttall). Office-furniture maker Steelcase owned a majority stake in the firm, but began divesting its shares through a five-year management buy-back program in 2007. The founders of the predecessor companies are still involved in the firm. The current CEO is Tim Brown.

Bill Moggridge died on September 8, 2012 during the development of IDEO. While the company started with a focus on designing consumer products (e. g. toothbrush, personal assistant, computers) by 2001, IDEO began to increase focus on consumer experiences (e. g. non-traditional classrooms). Kelley applied the term "design thinking" to business in order to encompass the approach to work of IDEO across industries and challenges.

In 2011, IDEO incubated (孵化) IDEO. org—a registered 501(c) (3) nonprofit dedicated to applying human-centered design to alleviate(减轻) poverty.

Acquisitions and partnerships

On October 17, 2017, IDEO acquired Datascope—a data science firm based in Chicago. Datascope has worked with IDEO as a consultant on many projects over the past four years. CEO Tim Brown states that the acquisition is largely motivated by advances in data sciences and machine learning. These advances allow for a bigger focus in human-centered applications including facilitation of the design process. Datascope's 15-person team will be moved to IDEO's Chicago office.

Products and services

IDEO has worked on projects in the consumer food and beverage, retail, computer, medical, educational, furniture, toy, office, and automotive industries. Some examples include Apple's first mouse, the Palm V PDA, and Steelcase's Leap chair. Clients include Air New Zealand, Coca-Cola, ConAgra Foods, Eli Lilly, Ford, Medtronic, Sealy, and Steelcase among many others.

The Apple Mouse   Palm V

OpenIDEO

In August 2010, IDEO introduced OpenIDEO — a collaborative platform for the design process. OpenIDEO was designed to be an internal tool for IDEO to collaborate with clients, but it is now a public tool. The purpose of the tool is to virtually drive the creative process to solve social problems, allowing for people of different expertise and backgrounds to collaborate. Examples of projects that have been facilitated by OpenIDEO include various projects of the WWF and TED Prize winner Jamie Oliver's Food Revolution movement.

OpenIDEO is an example of a crowdsourcing platform(众包平台) for creative work. On the website, one user puts up a brief summary of its idea and creates discussions. From the website, users can get help from others who have the knowledge or ability to provide him/her with helpful information or insightful guidance to develop the idea or that workpiece. There are three major steps for crowdsourcing on OpenIDEO: challenge, events, and alliances. The "Challenges" part is an open idea accelerator where worldwide users are connected to the platform. The "Events" part creates opportunities for users to participate, engage actively, and collaborate with innovators. The "Alliances" part provides a sense of community that supports its users to connect, design solutions together, and build partnerships.

2. Designworks

Designworks is a global creative consultancy owned by BMW and based in Newbury Park, California, United States. Designworks has two further studios, in Munich, Germany and Shanghai, China. Established independently in 1972 by Charles Pelly, it was acquired by BMW in 1995 and is now a wholly owned subsidiary (子公司) of BMW Group. The current president is Holger Hampf. The company has designed items such as the U. S. Olympic Team's bobsleds (雪橇) and Singapore Airlines' first class cabins.

It is a BMW subsidiary which was established by Charles Pelly, who brought DesignworksUSA from a start-up operation in his garage and made it one of the top ten consultancies in the world. Headquartered in the city of Newbury Park, California,

DesignworksUSA also operates an office in Munich, Germany. It became a wholly-owned BMW Group subsidiary in 1995, and was instrumental in the design of the BMW XL Sports Activity Vehicle and the BMW 5 Series. Other development projects at DreamWorks have included the BMW electric car, BMW 850 seat, BMW 3 Series E46, BMW Z8, BMW 7 Series interior, BMW Zeta show car, BMW 3 Series E46, and BMW 100 and 1200 Touring motorcycles, among many others. It is one of three BMW-owned facilities in California.

Designworks Headquarters

Origin

The design studio was founded in 1972 by designer Charles Pelly with the name DesignworksUSA (renamed Designworks in 2015). The company began in Malibu Canyon with three designers and early customers included Hyster and the Otis Elevator Company.

In 1978, the company expanded and moved to Van Nuys, and set up its first sister studio, D2, in Detroit. In 1986, Designworks designed the seats for the BMW 8 Series (E31), which was the company's first involvement with BMW. In the same year, the company moved to Agoura. In 1988, Designworks moved to Newbury Park, and began designing for Nokia and Siemens.

BMW ownership

In 1991, BMW acquired a large percentage of the company, and Designworks started its first car exterior design project in 1993 for the 1998 BMW 3 Series (E46). In May 1995, BMW purchased the remaining percentage of Designworks. At this time artist David Hockney painted an Art Car in the studio. In 1998, Designworks opened a studio in Munich, and in 1999, Henrik Fisker was named president.

In 2001, Adrian van Hooydonk became the president. In 2002, Designworks expanded its automotive studio in Newbury Park.

Verena Kloos was president from 2004 to 2009. In January 2006, the company opened a new

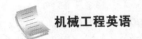

studio in Singapore, and in 2007 unveiled their new studio wing in Newbury Park. On December 1, 2009, Laurenz Schaffer, formerly director of Designworks' Munich studio, became the president. In April 2012, Designworks opened a new studio in Shanghai.

From 1 August 2016 Oliver Heilmer was appointed as the new president of Designworks. He remained in the role until 2017, when he moved to Munich to take over responsibility of MINI Design. The current president of Designworks is Holger Hampf. Appointed in the role on September 1, 2017, Holger originally joined Designworks in 1998, responsible for Product Design. From 2002 to 2010 he was a member of Designworks management team in Los Angeles.

Designworks has worked with a wide and varied range of clients including Embraer, Hewlett Packard, Microsoft, Starbucks, Coca-Cola, Boeing Business Jets, Pilatus, John Deere, Advanced Medical Optics, BAVARIA Yachts, Dornbracht, Acer, Head and Singapore Airlines.

3. frog

Frog (styled as frog) is a global design firm founded in 1969 by industrial designer Hartmut Esslinger in Mutlangen, Germany as "esslinger design". Soon after it moved to Altensteig, Germany, and then to Palo Alto, California, and ultimately to its current headquarters in San Francisco, California. The name was changed to Frogdesign in 1982 [the name apparently originating from an acronym (缩略词) for Esslinger's home country, the Federal Republic of Germany, originally printed lower-case (小写字母) as a rebellion against German grammatical rules which other companies adapted], then to Frog Design in 2000, and finally to frog in 2011.

In August 2004, the company announced that Flextronics International, a large electronics manufacturing services provider, was taking an equity stake (股权收益) in the company, a deal characterized by some commentators as essentially an acquisition. Flextronics CEO Michael Marks, in a March 2005 BusinessWeek article, said that Flex was going to integrate their San Jose-based industrial-design group with frog. The company is now a unit of Aricent (formerly Flextronics Software), which in turn is controlled by investment firm Kohlberg Kravis Roberts.

First designs were for WEGA in 1969, a German radio and television manufacturer, later acquired by Sony. frog continued to work for Sony and designed the Trinitron television receiver (电视接收器) in 1975. Their first designs for computer manufacturers were for proprietary systems (专有系统) by CTM (Computertechnik Müller) in 1970 and Diehl Data Systems in 1979. More prominent are the designs for Apple Computer, starting with the case of the portable Apple IIc, introducing the Snow White design language used by Apple during 1984—1990, and continuing with several Macintosh models. The firm designed Sun's SPARC stations in 1989 and the NeXT Computer in 1987.

Apple Ⅱc(ca. 1988)

4. Hatch Ltd

Hatch is a global multidisciplinary management, engineering and development consultancy. Its group companies have more than 9,000 staff in 70+ offices on six continents. In 2015, Hatch was ranked as a top 20 International Design Firm according to the Engineering News-Record(ENR,美国工程新闻记录).

The company was founded in Toronto, Ontario, Canada, by W. S. Atkins as W. S. Atkins & Associates in 1955. The company initially was involved in subway tunneling and other civil engineering projects, and expanded into metallurgy(冶金行业) when Gerry Hatch joined the company in January 1958. It became known as Hatch in 1962.

Hatch counts among its metals clients the top 20 (as measured by market capitalization) mining and metals companies in the world, including Alcan, Alcoa, BHP, Barrick Gold, BlueScope, Xstrata (formerly Falconbridge), Vale (formerly Inco) and Rio Tinto.

HATCH

Hatch Ltd Corporate Logo, 2016

In 1996 the company began an expansion program by purchasing several aligned engineering companies including Billiton Engineering (1996), Rescan Mining (1998), BHP Engineering (1999), Kaiser Engineers (2000), Acres International (2004), and MEK Engenharia (2012). By 2005 the combined billing of the company was around CDN $700 million.

Hatch today provides consulting, operations support, technologies, process design, and project and construction management to clients in three principal sectors: mining and metals; energy; and infrastructure.

The company's main offices are in Canada, Australia, South Africa, Chile, China, Brazil, Peru, Russia, United Kingdom, and United States. They also have several smaller offices

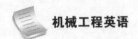

around the globe.

Its main business areas include:

- Mining & Mineral Processing—planning, design and maintenance consulting
- Energy—hydroelectric power, wind power, alternative energies (ocean, solar, geothermal, etc.), hybrid power (renewable and thermal), thermal power, nuclear power, oil and gas, and electricity transmission and distribution
- Infrastructure—light-rail transit (轻轨交通), freight/industrial rail (工业铁路), tunnels, water/wastewater (municipal/industrial), aviation, highways and bridges, ports and marine terminals, urban solutions
- Non-ferrous metals—nickel (镍), copper, zinc (锌), gold, platinum (铂), lithium (锂)
- Coal
- Light metals—aluminum and magnesium (镁)
- Iron & Steel—iron making, steel making, rolling and finishing
- Industrial minerals—titania slag (氧化钛渣), diamonds, potash(钾), phosphate (磷酸盐)
- Technologies—custom designed equipment and technology including autoclave (高压釜) for extractive metallurgy (萃取冶金), coilbox (卷取箱) for steel rolling(轧钢), electric arc furnace(电弧炉), fluidized bed (流化床), non-destructive testing

1. *Read the passage quickly by using the skills of skimming and scanning, and choose the best answer to the following questions.*

   1) In the first years of IDEO, who held a majority of stake in the firm but began divesting its shares in 2007? _____

      A. David Kelley   B. Bill Moggridge

      C. Mike Nuttall   D. Office-furniture maker Steelcase

   2) Before being acquired by IDEO, how long had Datascope been working with IDEO as a consultant? _____

      A. for about one year   B. for two years

      C. for three years   D. for four years

   3) About OpenIDEO, which of the following statements is NOT true? _____

      A. It was designed to be a public tool for worldwide users to collaborate with each other.

      B. Its purpose is to virtually drive the creative design process to solve social problems.

      C. It allows for people of different expertise and backgrounds to collaborate.

      D. It is a good example of a crowdsourcing platform for creative work.

   4) In _____ Designworks was acquired by BMW and became a subsidiary of BMW group.

      A. 1991   B. 1993

      C. 1995   D. 1986

5）Which was Designworks' first involvement with BMW? _____

    A. the design of BMW XL Sport Activity Vehicle in 1995

    B. the design of the BMW 5 Series in 1995

    C. the seat design of the BMW 8 Series in 1986

    D. the exterior design of BMW 3 Series in 1993

6）In 2017 _____ became the new president of Designworks.

    A. Verena Kloos   B. Holger Hampf

    C. Oliver Heilmer   D. Laurenz Schaffer

7）The current headquarters of frog is in _____.

    A. Mutlangen, Germany   B. Altensteig, Germany

    C. Plo Alto, California, the U. S.   D. San Francisco, California, the U. S.

8）Which of the following is NOT the company that frog once worked for in the 20th century? _____

    A. Flextronics   B. Sony

    C. Computertechnik Müller   D. Apple Computer

9）Which of the following statements about Hatch Ltd. is TRUE? _____

    A. Its group companies employ more than 1,000 staff on six continents.

    B. It was originally involved in metallurgy and some civil engineering projects, and expanded into subway tunneling later.

    C. In 1996 it began an expansion program by purchasing several aligned engineering companies.

    D. Its scope of business is mainly in three principal sectors: mining and metal, aeronautical engineering, and infrastructure.

10）Which company was ranked as a top 20 International Design Firm in 2015 according to the ENR? _____

    A. IDEO   B. Designworks

    C. frog   D. Hatch Ltd

2. *In this part, the students are required to make an oral presentation on either of the following topics.*

1）the development trend of industrial design

2）lessons from the success of IDEO/Designworks/frog/Hatch

习题答案

Unit Six Automobile Engineering

I. Pre-class Activity

Directions: *Please read the general introduction about* **Ferry Porsche** *and tell something more about the great scientist to your classmates.*

Ferdinand Anton Ernst Porsche (1909—1998), mainly known as Ferry Porsche, was an Austrian technical automobile designer and automaker-entrepreneur. He operated Porsche AG in Stuttgart, Germany. His father, Ferdinand Porsche, Sr. was also a renowned automobile engineer and founder of Volkswagen and Porsche. His nephew, Dr. Ferdinand Piëch, is the longtime chairman of Volkswagen Group, and his son, Ferdinand Alexander Porsche, was involved in the design of the Porsche 911.

Ferry Porsche's life was intimately connected with that of his father, Ferdinand Porsche, Sr., who began sharing his knowledge of mechanical engineering already in his childhood. With his father he opened a bureau of automobile design, in Stuttgart in 1931.

They worked together to fulfill their country's National Socialist regime's needs and they met Adolf Hitler at many business events. The Volkswagen Beetle was designed by Ferdinand Porsche, Sr. and a team of engineers, including Ferry Porsche.

After World War II, while his father remained imprisoned in France being accused of war crimes, Ferry Porsche ran their company. Aided by the postwar Volkswagen enterprise, he created the first cars that were uniquely associated with the company. Despite the political-economical adversities of the postwar years, the company manufactured automobiles, and eventually, became a world powerhouse for producing sports cars.

II. Specialized Terms

Directions：*Please remember the following specialized terms before the class so that you will be able to better cope with the coming tasks.*

A/C condenser 空调,冷凝器

air bag(安全)气囊

air-conditioning system 空调系统

alternative energy 替代能源

alternator（components）交流发电机

automatic transmission oil 自动变速箱油

battery n.电瓶

battery charger 电池充电器

bearing n.轴承

blower assembly 鼓风机

bolt n. 螺栓

brake n. 制动器,闸;刹车 vt. & vi. 刹(车)

bumper n. 减震器;保险杠

bumper pad 防撞护垫

buzzer n. 蜂鸣器

car audio 汽车音响

car body design 车身设计

car burglar alarm 防盗器

car CD 汽车用光盘

car hand-free mobile phone 汽车用行动电话

car LCD 汽车用液晶显示器

car navigation system 汽车导航系统

car security system 汽车保全系统

casting parts(processing)铸造件(加工)

central door lock 中央门控

chassis n. (车辆的)底盘

clutch vt. 抓紧;紧握 vi. 踩离合器 n. 紧抓;控制;离合器

clutch master cylinder 离合器总泵

coaster brake 脚刹车器

combination meter 组合仪表

console n. 中央置物箱

cooling system 冷却系统

cosmetics for automobile 汽车清洁保养用品

cruise controller 定速器

cylinder n. 圆筒,圆柱;汽缸

derailleur n. 变速器

distributor n. 分电盘

door armrest 车门扶手

door handle 车门把手

door hinge 门铰链

door mirror 后视镜

electric seat 电动座椅

electrical parts 电装品

engine hood 引擎盖

exhaust pipe 排气管

exterior parts 外装品

flasher n. 闪光器

flywheel 飞轮

forging parts(processing)锻造件(加工)

fuel gauge 油量表

fuel pipe 燃油管

fuel tank 油箱

fueling system 燃料系统

gasoline-powered car 汽油动力车

gear n. 齿轮

guard assy 防撞杆

headlining n. 顶篷

heat insulator 隔热、保温材料,热绝缘器

hinge of engine hood 引擎盖铰链

hydraulic brake 液压式刹车器

ignition cable 高压线组

ignition coil 点火线圈

ignition coil module 点火线圈模块

install vt. 安装；安置

instrument panel 仪表板

interior parts 内装品

lighting controller 车灯控制器

lubrication system 润滑系统

luxury car 豪华车

maintenance n. 维持；保养；维护；维修

monocoque body 单体车身

motor components 马达零件

moulding n. 压条

mud guard 挡泥板

oil filter for automatic transmission 自排车用滤油器

pedal n. 踏板

piston n. 活塞

power seat unit 电动座椅装置

production, test & painting equipment 生产、检测及涂装设备

propeller shaft 螺旋轴，传动轴

radiator n. 水箱

rear view display 倒车显示器

regulator n. 调整器

reverse sensor 倒车雷达

rubber parts 橡胶件

satellite navigation system 卫星导航系统

seat belt 安全带

shift lever 变速杆

shock absorber 避震器

side protector 车身护条

stamping parts 钣金件

steering linkage 转向连杆

steering wheel 方向盘

sun roof 天窗

SUV abbr. *sport utility vehicle* 运动型多功能车

transmission n. 传动装置，变速器

transmission box 变速箱

wall charger 壁式充电器

wheel cover 轮圈盖

wheel disk 轮圈

wheeling system 车轮系统

window lifter handle 车窗升降摇柄

windshield n. 挡风玻璃

wiper/ linkage 雨刷及雨刷连杆

III. Watching and Listening

Task One　Charging Ahead: the Case for Plug-in Hybrid Cars（I）

视频链接及文本

New Words

addicted adj. 上瘾的；沉迷于

headlong adv. 头向前地；急速地　adj. 头向前的；急速的

collision n. 碰撞，冲突

superpower n. 超级大国

transition n. 过渡，转变

petroleum n. 石油

bio-fuel 生物燃料

in a nutshell 简言之

hybrid n. 混合动力汽车　adj. 混合的

mileage n. 英里数，里程

plug-in n. 插件程序

booming adj. 急速发展的

grid n. 格子，系统网络

cusp n. 尖头，尖端

gallon n. 加仑(容量单位)

emission n. 排放,辐射;排放物,散发物
  (尤指气体)

conversion n. 变换,转变 auto-maker 汽车
  制造商

modification n. 修改,改良

drive train 传动系统

hurdle n. 障碍,困难

lithium ion 锂离子(电池)

lifespan n. 寿命;使用期

acceleration n. 加速

tackle vt. 着手处理

disposal n. 处置;清理

commute vi. 通勤 n. 通勤来往

Exercises

1. *Watch the video for the first time and choose the best answer to the following questions.*

   1) According to the video, which of the following is NOT a problem caused by Americans' love for cars? _____

      A. pollution   B. global warming

      C. high fuel costs   D. waste of resources

   2) Today's hybrid car, like the Prius, uses a combination of _____ as its fuel.

      A. petroleum and gas   B. bio-fuels and petroleum

      C. electricity and gas   D. electricity and bio-fuels

   3) Several factors promote today's hybrid to the plug-in hybrid. Which of the following is NOT one of them? _____

      A. the finding of alternative fuels

      B. the booming popularity of hybrid vehicles

      C. the increased energy efficiency

      D. development of the nation's electricity grid

   4) According to the video, every plug-in hybrid with the new powerful battery added to it will save _____ gasoline.

      A. 80 percent   B. 70 percent

      C. 60 percent   D. 85 percent

   5) As for the batteries of the plug-in hybrids, which of the following is NOT true? _____

      A. Right now the high cost of the battery is the major barrier to be overcome.

      B. The battery only needs to last two or three years.

      C. It's believed that the costs will come down with mass production.

      D. The batteries need to be powerful and have a long lifespan.

2. *Watch the video again and decide whether the following statements are true or false.*

   1) The United States is a country that is self-sufficient in oil. (   )

   2) The demand for energy in the world has caused a lot of troubles and collisions among countries. (   )

   3) Since the regular hybrids are not good enough, the Americans are trying to take the technology one step further to contrive the plug-in hybrid. (   )

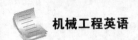

4) To convert Prius into plug-in hybrids, some modifications are to be made to the battery and the drive train of the car. (   )

5) Beside the battery, safety and disposal pose challenges to engineers. (   )

3. *Watch the video for the third time and fill in the following blanks.*

Today most plug-in hybrids are _____ of existing hybrids, mainly the Prius. The fact that you can't buy plug-ins right now is a problem, because we believe that plug-in hybrids should be commercially _____. So, in order to do that, we actually converted some Priuses into plug-in hybrids so that we didn't have to wait for the _____ to make them. The main _____ are in the back of the car. We added a larger _____. So if so many people believe the plug-in hybrid is such a great idea, why don't we have them already? Many of the reasons _____ to one item. So you can say the big problem is _____ and the big problem with the batteries is cost. So that's the major _____ that has to be overcome. To _____ consumers, the battery needs to be powerful and have a long lifespan. Other issues being _____ by engineers concern safety and disposal.

4. *Share your opinions with your partners on the following topics for discussion.*

1) What are the major problems today's hybrids are faced with? What do engineers need to do to solve them?

2) Do you think the prospect of the plug-in hybrid cars is promising or not? Why?

Task Two   Charging Ahead: the Case for Plug-in Hybrid Cars (Ⅱ)

视频链接及文本

New Words

roadblock n. 路障

viable adj. 切实可行的;有望实现的

kilowatt n. [电]千瓦

pencil out 削减开支、费用等

offset vt. 抵消;补偿

utility n. 公用;公用事业公司

incentive n. 动机;诱因

infrastructure n. 基础设施;基础建设

carport n. <美>车库,车棚

literally adv. 照字面地; 确实地,真正地

humanity n. 人类;人性

cataclysm n. (突然降临的)大灾难

vulnerable adj. (地方)易受攻击的;易受伤的

grip n. /v. 紧握,抓牢

Exercises

1. *Watch the video for the first time and choose the best answer to the following questions.*

1) The Americans had an electric and hybrid vehicle program _____.

    A. in 1976　　　　　　　　B. in Detroit

    C. launched by google. org　　D. supported by the government

2) From the video we can know that the smart grid will benefit us in a few ways EXCEPT _____.

    A. it would allow consumers to decide how much they spend on electricity

B. consumers could make choices about when they use electricity

    C. it could make better use of alternative energy

    D. consumers could decide the price of power

3) A three-kilowatt solar system installed for a house _____.

    A. is economical

    B. could save the electrical use for the house

    C. reduce gasoline use for the hybrid car

    D. all of the above

4) The plug-in hybrid has an advantage over other alternative. It's that _____.

    A. it can be recharged faster

    B. it requires little change to the nation's infrastructure

    C. it is a technology that has just about arrived

    D. it could solve all the problems caused by energy shortages

5) We can learn from the video that _____.

    A. every household may have the plug-in hybrid vehicle

    B. plug-in hybrids will gain a larger part of market share in the near future

    C. with few limitations, the plug-in hybrid is an ideal vehicle

    D. it's believed that plug-in hybrids are one of the technologies that can change the world

2. *Watch the video again and decide whether the following statements are true or false.*

    1) It is difficult to convince the executives of automakers of the promising future of electric and vehicles. (　)

    2) A few auto makers are promising to manufacture plug-in hybrids in a few years. (　)

    3) The plug-in hybrid car could not be recharged by the smart grid but be recharged with energy from the sun. (　)

    4) The auto industry, the utility companies and the government are all involved in raising this country's level of energy efficiency. (　)

    5) As long as you live in the suburbs, you can use a plug-in hybrid. (　)

3. *Watch the video for the third time and fill in the following blanks.*

    The auto industry is beginning to see that these cars are _____, and at least a few companies are promising _____ of plug-in hybrids in a few years. So what we are going to try to help that market is to put some _____ on the table for _____ consumers that will buy a plug-in car, for utilities that will help build out this _____, and for the auto manufacturers who will actually make these _____ cars.

    Many people believe plug-in hybrids will provide _____ for our society, and that they are even _____ for our future. It is believed that this plug-in hybrid area may be critical to the _____ of humanity. It is clear that all roads lead to alternative fuel for our cars. Many scientists believe that if we don't change _____ soon, there will be serious

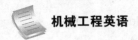

consequences.

4. *Share your opinions with your partners on the following topics for discussion.*

1) If you were the CEO of an auto company, would you like to produce this plug-in hybrid? Why?

2) Suppose you were a salesperson in an automotive 4S store, how would you introduce a hybrid car to a customer?

IV. Talking

Task One   Classical Sentences

Directions: *In this section, some popular sentences are supplied for you to read and to memorize. Then, you are required to simulate and produce your own sentences with reference to the structure.*

General Sentences

1. If it doesn't rain tomorrow, I think I'll go shopping.
   如果明天不下雨,我想去购物。

2. There's a possibility we'll go, but it all depends on the weather.
   我们有可能去,但要看天气怎么样。

3. Let's make a date to go shopping next Thursday.
   我们约好下周四去购物吧。

4. If I have time tomorrow, I think I'll get a haircut.
   如果我明天有空,我想去剪头发。

5. I hope I remember to ask the barber not to cut my hair too short.
   我希望我记得叫理发师不要把我的头发剪得太短。

6. If I get my work finished in time, I'll leave for New York Monday.
   如果我能及时完成工作,我周一就去纽约。

7. Suppose you couldn't go on the trip, how would you feel?
   设想一下,如果你不能去旅游,你会有什么感觉?

8. What would you say if I told you I couldn't go with you?
   如果我告诉你我不能和你一起去,你会怎么想?

9. If I buy the car, I'll have to borrow some money.
   如果我想买那辆车的话,我就得借些钱。

10. We may be able to help you in some way.
    我们也许能在某些方面帮助你。

11. If you were to attend the banquet, what would you wear?
    如果你要参加宴会,你会穿什么?

12. What would you have done last night if you hadn't had to study?
    如果昨天晚上你不用学习的话,你会做什么?

13. I would have gone on the picnic if it hadn't rained.
    要不是下雨,我就去野炊了。

14. If you had gotten up earlier, you would have had time for breakfast.
    如果你早一点起床,就有时间吃早饭了。

15. If I had had time, I would have called you.
    我要是有时间,就给你打电话了。

16. Would he have seen you if you hadn't waved to him?
    要是你没向他挥手,他还能看见你吗?

17. If he had only had enough money, he would have bought that house.
    他要是有足够的钱,就会买下那房子了。

18. I wish you had called me back the next day, as I had asked you to.
    可惜你没有按我的要求,在第二天给我回个电话。

19. If you hadn't slipped and fallen, you wouldn't have broken your leg.
    如果你没滑跌倒,你就不会摔断了腿。

20. If I have known you want to go, I would have called you.
    要是我知道你想去,我就叫你了。

21. Had I known you didn't have the key, I wouldn't have locked the door.
    要是我知道你没有钥匙,我就不会锁门了。

22. She would have gone with me, but she didn't have time.
    她本想和我一起去的,可是她没时间。

23. If I had asked directions, I wouldn't have got lost.
    要是我问一下路,就不会走丢了。

24. Even if we could have taken the vocation, we mightn't have wanted to.
    即使我们可以休假,我们也许不想去呢。

25. Everything would be alright, if you had said that.
    如果你是那样说的,一切都好办了。

26. Looking back on it, I wish we hadn't been given in so easily.
    现在回想起来,我真希望我们没有那么轻易地让步。

27. One of these days, I'd like to take a vacation.
    总有一天,我要去休假。

28. As soon as I can, I'm going to change jobs.
    我要尽快换个工作。

29. There's a chance he won't be able to be home for Christmas.
    他可能不能回家过圣诞节了。

30. What is it you don't like the winter weather?
    你为什么不喜欢冬天的天气?

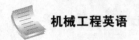

31. I don't like it when the weather gets really cold.
我不喜欢天太冷。

32. The thing I don't like about driving is all the traffic on the road.
我不喜欢开车去是因为路上很拥挤。

33. He doesn't like the idea of going to bed early.
他不喜欢早睡。

34. I like to play tennis, but I'm not a very good player.
我很喜欢打网球,但是打得不是很好。

35. I don't like spinach even though I know it's good for me.
我不喜欢菠菜,尽管我知道菠菜对我有好处。

36. I'm afraid you're being too particular about your food.
恐怕你对食物太挑剔了。

37. He always finds fault with everything.
无论是什么事情,他总能发现错误。

38. You have wonderful taste in clothes.
你对衣服很有品位。

39. What's your favorite pastime?
你最喜欢的消遣是什么?

40. What did you like best about the movie?
你最欣赏这部电影的哪个方面?

41. Will the new movie be welcomed by its spectators or just raise another ware of disappointment?
那么这部新的电影将会受到观众欢迎,还是引起新一波的失望呢?

42. The feature started at 9 o'clock and ended at 11:30.
专题片从九点开始,一直持续到十一点半。

43. They say the new film is an adventure story.
他们说这部新影片讲的是一个冒险故事。

44. We went to a concert last night to hear the symphony orchestra.
我们昨天晚上去听交响乐了。

45. A group of us went out to the theater last night.
昨晚我们一群人去了剧院。

46. The new play was good and everybody enjoyed it.
这个新剧很好,人人都喜欢。

47. By the time we got there, the play had already begun.
我们到达时,戏已经开始了。

48. The usher showed us to our seats.
引导员把我们带到了座位前。

49. The cast of the play included a famous actor.

这场戏的演员阵容里有一位很著名的演员。

50. After the play was over, we all wanted to get something to eat.
戏结束了以后,我们都想去吃点东西。

Specialized Sentences

1. They're using low-polluting bio-diesel fuel in all their trucks and heavy equipment.
他们在所有卡车和重型机械中都使用低污染的生物燃料。

2. The functional design and development of a modern motor vehicle is typically done by a large team from many different disciplines.
现代汽车的功能设计和开发通常由来自许多不同学科的大型团队完成。

3. Automotive design as a professional vocation is practiced by designers who may have an art background and a degree in industrial design or transportation design.
汽车设计是一种专业化的职业,从事该职业的设计师应具有艺术背景,以及工业设计或交通设计学位。

4. Automotive design focuses not only on the isolated outer shape of automobile parts, but concentrates on the combination of form and function.
汽车设计不仅注重孤立的汽车零部件外形,而且注重形式与功能的结合。

5. The aesthetic value will need to correspond to ergonomic functionality and utility features as well.
审美价值也需要与人体工程学的功能和实用功能相符合。

6. Automobile interior design is first done by a series of digital or manual drawings.
汽车内部设计首先是通过一系列的数字或手工绘图完成的。

7. The clay model is still the most important tool to evaluate the design of a car.
黏土模型仍然是评价汽车设计最重要的工具。

8. The design team also develops graphics for items such as badges, decals, dials, switches, and so on.
设计团队还为一些物品开发图形,如徽章、贴花、刻度盘、开关,等等。

9. During the development process succeeding phases will require the 3D model fully developed to meet the aesthetic requirements of a designer as well as all engineering and manufacturing requirements.
在开发过程中,后续阶段将需要完全开发 3D 模型,以满足设计师的审美需求以及所有工程和制造要求。

10. The fully developed CAS digital model will be redeveloped for manufacturing.
完全开发的 CAS 数字模型将被重新开发用于制造。

11. Integration of an automobile involves fitting together separate parts to form a monocoque body or units and mounting these onto a frame, the chassis.
汽车的集成包括将不同的部件组装在一起形成单体车体或部件,然后将这些部件安装到车架(底盘)上。

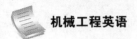

12. The chassis is usually tested on the road before the complete body of the vehicle is attached.

底盘通常先在道路上测试,然后才将整个车身组装起来。

13. Most of the automakers were more concerned with mechanical reliability rather than with its external appearance.

大多数汽车制造商更关心机械的可靠性,而不是它的外观。

14. BMW entered the automobile design with a sporty looking.

宝马以运动型外观进入汽车设计领域。

15. The styling team for a specific model consists of a chief designer and an exterior as well as interior designer.

特定模型的造型团队由首席设计师、外观设计师以及内饰设计师组成。

16. It would not be possible for automobiles to meet modern safety and fuel economy requirements without electronic controls.

没有电子控制装置,汽车将无法满足现代安全性和燃油经济性的要求。

17. High quality is needed to meet customer requirements and to avoid expensive recall campaigns.

为了满足客户的要求,并避免昂贵的召回活动,需要有高质量的产品。

18. The complexity of components involved in the production process requires a combination of different tools and techniques for quality control.

生产过程中涉及的组件的复杂性要求结合不同的工具和技术进行质量控制。

19. Hybrids, unlike pure electric cars, never need to be plugged in.

与纯电动汽车不同,混合动力汽车不需要外接电源充电。

20. He is repairing the brake lever of an automobile.

他正在修理汽车的刹车杆。

21. This car has done 2,000 mileages today.

这辆车今天行驶了 2,000 英里。

22. They only pay for the excess power that flows into the utility's power grid.

他们只为流入电力公司电网的多余电力付费。

23. I install a hub dynamo into a front wheel.

我把一个轮毂发电机装到了前车轮上。

24. The company serves the national automotive aftermarket with a broad range of accessory and recreational-vehicle products.

这家公司为全国汽车售后市场提供各种零配件和休闲旅游车产品。

25. General Motors Corp. is the world's largest automaker.

通用汽车公司是世界上最大的汽车制造商。

26. The drive line or drive train transfers the power of the engine to the wheels.

传动链或传动系将发动机的动力传递给车轮。

27. The car will retain the typical BMW platform and drive train for front engine and rear

wheel drive.

这款车将保留典型的宝马平台以及传动系统的前置引擎和后轮驱动。

28. Acceleration to 60 mph takes a mere 5. 7 seconds.

加速到时速 60 英里只需要 5. 7 秒。

29. Cars could be plugged into walls to recharge.

汽车可以插入墙壁充电。

30. You can buy a USB battery charger almost anywhere that sells rechargeable batteries.

你能在大多数卖充电电池的地方买到 USB 电池充电器。

31. This car is equipped with an in-car use battery charger for drive out use.

这款汽车配备有一个车载电池充电器,用于驾车外出时使用。

32. The brake doesn't grip properly.

刹车不灵。

33. Use the brake gently or you'll stall the engine.

轻踩刹车,不然的话发动机会熄火的。

34. The front bumper is integrated into the design, further softening the look.

前保险杠融入整体设计中, 进一步改善了外观。

35. We have to stop using so much oil, and that means we must find alternative fuels.

我们必须停止使用这么多的石油,这意味着我们必须找到替代燃料。

36. We're standing on the cusp of a new generation of vehicles that can truly revolutionize our national energy paradigm.

我们正站在新一代汽车的尖端,这些汽车将真正改变我们国家的能源模式。

37. Today's hybrid cars, like the Prius, use a combination of electricity and gas.

今天的混合动力汽车,正如普锐斯,使用电力和天然气的组合获得动力。

38. Hybrid cars still rely on gas to power the engine and to charge the battery.

混合动力汽车仍然需要汽油来驱动引擎和给电池充电。

39. It's believed costs will come down with mass production.

人们相信大规模生产会降低成本。

40. To appeal to consumers, the battery needs to be powerful and has a long lifespan.

为了吸引消费者,电池必须功能强大且使用寿命长。

41. Other issues being tackled by engineers concern safety and disposal of the battery.

工程师正在处理的其他问题涉及电池的安全和处置。

42. The auto industry is beginning to see that hybrid cars are in demand.

汽车行业开始看到混合动力汽车的市场需求。

43. The smart grid could make better use of alternative energy.

智能电网可以更好地利用替代能源。

44. During the development of a new car, the design of car body is very important and will influence the performance of the product.

在汽车新车型设计开发过程中,汽车车身设计非常重要,且将影响到产品的性能。

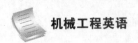

45. I would like to rent a car with a satellite navigation system.

我想要租一辆有卫星导航系统的车。

46. Now, four-cylinder cars are no longer produced.

现在,四汽缸汽车已不再生产。

47. The nonmetal material engine hood can be opened completely by hand, convenient for the engine maintenance.

非金属材料发动机罩可手动完全打开,便于发动机的维修。

48. The fuel gauge indicated that the tank was quite full.

燃油表显示,油箱已满。

49. As a very important part of the ignition system, ignition coil gives great impact to the performance of motorcars.

点火线圈作为汽车点火控制系统的重要组成部件,对汽车的性能影响很大。

50. This mud guard board applies to Honda, quality and cheap.

这款挡泥板适用于本田车系,质优价廉。

Task Two   Sample Dialogue

Directions:*In this dialogue, you are going to read several times the following sample dialogue about the relevant topic. Please pay special attention to five Cs (culture, context, coherence, cohesion and critique) in the dialogue and get ready for a smooth communication in the coming task.*

(*in an automobile 4S shop*)

A: Good morning, sir. Can I help you?

B: I'd like to buy an automatic car with a price of about 50,000 dollars.

A: I think you've made the right choice, coming to us. We have a wide selection of vehicles you can choose from. I'll show you. This way please.

B: And I'd like a car with a good stereo.

A: Please rest assured. All of our cars have excellent stereos in them. Stereos, air-conditioning, they are all of high standard.

B: This is a Porsche!

A: Yes. Beautiful, isn't it?

B: It must be really expensive. I can't afford it.

A: How about this Ford? It's a high-class vehicle, it's safe, stable, high-performance and in a nice shape. Do you like its exterior?

B: Not bad. May I have a test drive?

A: Of course! Take a test drive and you'll quickly understand that this is something completely different.

B: I'm glad it is an automatic. I don't like having to change gears.

A: Yes, automatics are very simple to drive. Put your seat belt on, check the side mirrors and the rear view mirror, check the petrol situation. Now put the keys in the ignition and start it up.

(**several minutes later**)

A: How are you feeling now?

B: Fantastic! How about the after-service?

A: Definitely we provide great lifetime after-sales service. And, to keep the car in good mechanical condition, we usually recommend that you bring your car to have a tune up in every five thousand kilometers. Trust me, regular tune ups will keep your car running smoothly and avoid break downs.

B: OK. Thank you. I will consider about it seriously.

Task Three   Simulation and Reproduction

Directions: *The class will be divided into three major groups, each of which will be assigned a topic. In the process of discussion, please observe the principles of cooperation, politeness and choice of words. One of the groups will be chosen to demonstrate the discussion to the class.*

1) the main challenges facing the auto industry

2) the future trend of automobile development

3) the influences of automobiles on our life

Task Four   Discussion and Debate

Directions: *The class will be divided into two groups. Please choose your stand in regard to the following controversy and support your opinions with scientific evidences. Please refer to the specialized terms and classical sentences in the previous parts of this unit.*

Since the beginning of the 21st century, technological development has realized the democratization of automobiles. Economical vehicles are very popular worldwide and the automakers across the world compete with each other by introducing their low-priced cars. In the meantime, the luxury cars and entry-level luxury cars are more favored by rich people. Some famous auto companies try to produce high-priced luxury vehicles. To maintain its market share and brand image, an automaker may adopt both strategies. Which strategy is better? Please give your reasons.

V. After-class Exercises

1. *Match the English words in Column A with the Chinese meaning in Column B.*

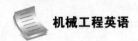

	A		B
1)	battery charger	A)	离合器
2)	regulator	B)	踏板
3)	clutch	C)	减震器;保险杠
4)	radiator	D)	调整器
5)	chassis	E)	电池充电器
6)	pedal	F)	顶篷
7)	transmission	G)	水箱
8)	headlining	H)	(车辆的)底盘
9)	bumper	I)	汽缸
10)	cylinder	J)	传动装置

2. *Fill in the following blanks with the words in the word bank. Change the form if it's necessary.*

hybrid	battery	ignition	lubrication	chassis
install	acceleration	brake	clutch	wheel

1) It's a 35-kilowatt-hour lithium titanate(钛酸盐)_____ pack that is quickly rechargeable and safe.

2) A front-wheel-drive _____ would serve as the underpinning(支承结构) for many of Ford's front-wheel-drive vehicles.

3) Laura let out the _____ and pulled slowly away down the lane.

4) Hybrids have proven themselves especially well with regard to inexpensive _____ repairs, thanks to regenerative braking.

5) Stepping into his garage he opens the driver's door, sits inside, flicks on the _____ and listens as the four turbofans blast into life.

6) All _____ drive is an increasingly popular feature for luxury cars, and not just for trucks.

7) Of course, a _____ or fuel-efficient car emits less carbon dioxide than an SUV.

8) Grease gun and oiler are the _____ service tools for machinery.

9) The performance version boasts 0~60 _____ in less than five seconds, which few other SUVs can match.

10) Auto executives predicted that having to _____ seatbelts would bring the downfall of their industry.

3. *Find a word from Part II Specialized Terms which means*：

1) process of keeping something in good condition by regularly checking it and repairing it when necessary

2) a device in a vehicle that makes it go slower or stop

3) moved or operated or effected by liquid (water or oil)

4) the system of gears and shafts by which the power from the engine reaches and turns

the wheels

5) the act of guiding or showing the way

6) the act of grasping; a pedal or lever that engages or disengages a rotating shaft and a driving mechanism

7) put something somewhere so that it is ready to be used

8) a device for charging or recharging batteries

4. *Translate the following sentences into English.*

1) 重要的是,我们要继续实现从石油向生物燃料的转变。

_____

_____

2) 我们的最终目标是让汽车制造商相信插电式混合动力车是未来的发展方向。

_____

_____

3) 智能电网将允许消费者选择何时、花多少钱买电。

_____

_____

4) 能够真正改变世界的新技术很少出现。

_____

_____

5) 我们离高能源效率、低环境影响、低成本这个水平还有很远的路要走。

_____

_____

5. *Please write an essay of about 120 words on the topic*: ***My ideal private car.*** *Some specific examples will be highly appreciated and you have to watch out the spelling of some specialized terms you have learnt in this unit.*

_____

_____

_____

_____

_____

_____

_____

_____

_____

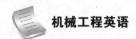

VI. Additional Reading

Ford Motor Company

Ford Motor Company is an American multinational automaker headquartered in Dearborn, Michigan, a suburb of Detroit. It was founded by Henry Ford and incorporated on June 16, 1903. The company sells automobiles and commercial vehicles under the Ford brand and most luxury cars under the Lincoln brand. Ford also owns Brazilian SUV manufacturer Troller, an 8% stake in Aston Martin of the United Kingdom, and a 49% stake in Jiangling Motors of China. It also has joint-ventures in China (Changan Ford), Taiwan (China Ford Lio Ho), Thailand (AutoAlliance Thailand), Turkey (Ford Otosan), and Russia (Ford Sollers). The company is listed on the New York Stock Exchange and is controlled by the Ford family; they have minority ownership but the majority of the voting power.

The Ford World Headquarters in Dearborn, Michigan

Ford introduced methods for large-scale manufacturing of cars and large-scale management of an industrial workforce using elaborately engineered manufacturing sequences typified by moving assembly lines; by 1914, these methods were known around the world as Fordism. Ford's former UK subsidiaries Jaguar(捷豹) and Land Rover(路虎), acquired in 1989 and 2000 respectively, were sold to Tata Motors(塔塔汽车) in March 2008. Ford owned the Swedish automaker Volvo from 1999 to 2010. In 2011, Ford discontinued the Mercury(水星) brand, under which it had marketed entry-level luxury cars in the United States, Canada, Mexico, and the Middle East since 1938.

Ford is the second-largest U. S. -based automaker (behind General Motors) and the fifth-largest in the world (behind Toyota, VW, Hyundai-Kia and General Motors) based on 2015 vehicle production. At the end of 2010, Ford was the fifth largest automaker in Europe. The company went public in 1956 but the Ford family, through special Class B shares, still retain 40 percent voting rights. During the financial crisis at the beginning of the 21st century, it was close to bankruptcy, but it has since returned to profitability. Ford was the eleventh-ranked overall American-based company in the 2018 Fortune 500 list, based on global revenues in 2017 of $156.7 billion. In 2008, Ford produced 5.532 million automobiles and employed about 213,000 employees at around 90 plants and facilities worldwide.

20th century

Henry Ford's first attempt at a car company under his own name was the Henry Ford Company on November 3, 1901, which became the Cadillac Motor Company on August 22, 1902, after Ford left with the rights to his name. The Ford Motor Company was launched in a converted factory in 1903 with $28,000 in cash from twelve investors, most notably John and Horace Dodge (who would later found their own car company). The first president was not Ford, but local banker John S. Gray, who was chosen to assuage(缓解) investors' fears that Ford would leave the new company the way he had left its predecessor. During its early years, the company produced just a few cars a day at its factory on Mack Avenue and later its factory on Piquette Avenue in Detroit, Michigan. Groups of two or three men worked on each car, assembling it from parts made mostly by supplier companies contracting for Ford. Within a decade, the company would lead the world in the expansion and refinement of the assembly line concept, and Ford soon brought much of the part production in-house in a vertical integration that seemed a better path for the era.

Henry Ford (ca. 1919)   A 1910 Model T, Photographed in Salt Lake City

Henry Ford was 39 years old when he founded the Ford Motor Company, which would go on to become one of the world's largest and most profitable companies. It has been in continuous family control for over 100 years and is one of the largest family-controlled companies in the world.

The first gasoline-powered automobile had been created in 1885 by the German inventor Carl Benz (Benz Patent-Motorwagen). More efficient production methods were needed to make automobiles affordable for the middle class, to which Ford contributed by, for instance, introducing the first moving assembly line in 1913 at the Ford factory in Highland Park.

Between 1903 and 1908, Ford produced the Models A, B, C, F, K, N, R, and S. Hundreds or a few thousand of most of these were sold per year. In 1908, Ford introduced the mass-produced Model T, which totaled millions sold over nearly 20 years. In 1927, Ford replaced the T with the Model A, the first car with safety glass in the windshield. Ford launched the first low-priced car with a V8 engine in 1932.

In an attempt to compete with General Motors' mid-priced Pontiac, Oldsmobile, and Buick, Ford created the Mercury in 1939 as a higher-priced companion car to Ford. Henry Ford purchased the Lincoln Motor Company in 1922, in order to compete with such brands as Cadillac and Packard for the luxury segment of the automobile market.

In 1929, Ford was contracted by the government of the Soviet Union to set up the Gorky Automobile Plant in Russia initially producing Ford Model A and AAs thereby playing an important role in the industrialization of that country.

The creation of a scientific laboratory in Dearborn, Michigan in 1951, doing unfettered(不受约束的) basic research, led to Ford's unlikely involvement in superconductivity(超导电性) research. In 1964, Ford Research Labs made a key breakthrough with the invention of a superconducting quantum interference device(量子干涉仪) or SQUID.

Ford offered the lifeguard safety package from 1956, which included such innovations as a standard deep-dish steering wheel, optional front, and, for the first time in a car, rear seatbelts, and an optional padded dash. Ford introduced child-proof(对儿童安全的) door locks into its products in 1957, and, in the same year, offered the first retractable hardtop on a mass-produced six-seat car.

In late 1955, Ford established the Continental division as a separate luxury car division. This division was responsible for the manufacture and sale of the famous Continental Mark II. At the same time, the Edsel division was created to design and market that car starting with the 1958 model year. Due to limited sales of the Continental and the Edsel disaster, Ford merged Lincoln, Mercury, and Edsel into "M-E-L", which reverted to "Lincoln-Mercury" after Edsel's November 1959 demise.

The Ford Mustang was introduced in April, 17, 1964 during New York World's Fair. In 1965, Ford introduced the seat belt reminder light.

With the 1980s, Ford introduced several highly successful vehicles around the world. During the 1980s, Ford began using the advertising slogan—"Have you driven a Ford, lately?" to introduce new customers to their brand and make their vehicles appear moremodern. In 1990 and 1994 respectively, Ford also acquired Jaguar Cars(捷豹汽车) and Aston Martin(阿斯顿·马丁). During the mid-to-late 1990s, Ford continued to sell large numbers of vehicles, in a booming American economy with a soaring stock market and low fuel prices.

With the dawn of the new century, legacy health care costs, higher fuel prices, and a faltering economy led to falling market shares, declining sales, and diminished profit margins. Most of the corporate profits came from financing consumer automobile loans through Ford Motor Credit Company.

21st century

By 2005, both Ford and GM's corporate bonds had been downgraded to junk status, as a result of high U. S. health care costs for an aging workforce, soaring gasoline prices, eroding market share, and an over dependence on declining SUV sales. Profit margins decreased on large vehicles due to increased "incentives" (in the form of rebates or low interest financing) to offset declining demand. In the latter half of 2005, Chairman Bill Ford asked newly appointed Ford Americas Division President Mark Fields to develop a plan to return the company to profitability. Fields previewed the Plan, named The Way Forward, at the December 7, 2005, board meeting of the company and it was unveiled to the public on January 23, 2006. "The Way Forward" included resizing the company to match market realities, dropping some unprofitable and inefficient models, consolidating production lines, closing 14 factories and cutting 30,000 jobs.

Ford moved to introduce a range of new vehicles, including "Crossover SUVs(跨界越野车)" built on unibody car platforms(整体式汽车平台), rather than more body-on-frame chassis(车体底盘). In developing the hybrid electric power train technologies for the Ford Escape Hybrid SUV, Ford licensed similar Toyota hybrid technologies to avoid patent infringements(侵权). Ford announced that it will team up with electricity supply company Southern California Edison (SCE) to examine the future of plug-in hybrids in terms of how home and vehicle energy systems will work with the electrical grid. Under the multimillion-dollar, multi-year project, Ford will convert a demonstration fleet of Ford Escape Hybrids into plug-in hybrids, and SCE will evaluate how the vehicles might interact with the home and the utility's electrical grid. Some of the vehicles will be evaluated "in typical customer settings", according to Ford.

William Clay Ford Jr. , great-grandson of Henry Ford (and better known by his nickname "Bill"), was appointed executive chairman in 1998, and also became chief executive officer of the company in 2001. Ford sold motorsport engineering company Cosworth to Gerald Forsythe and Kevin Kalkhoven in 2004, the start of a decrease in Ford's motorsport involvement. Upon the retirement of president and chief operations officer Jim Padilla in April 2006, Bill Ford assumed his roles as well. Five months later, in September, Ford named Alan Mulally as president and CEO, with Ford continuing as executive chairman. In December 2006, the company raised its borrowing capacity to about $25 billion, placing substantially all corporate assets as collateral. Chairman Bill Ford has stated that "bankruptcy is not an option". Ford and the United Auto Workers, representing approximately 46,000 hourly workers in North America, agreed to a historic contract settlement in November 2007 giving the company a substantial break in terms of its ongoing retiree health care(退休人员医疗保健) costs and other economic issues. The agreement included the establishment of a company-funded, independently run Voluntary Employee Beneficiary Association (VEBA) trust to shift the burden of retiree health care from the company's books, thereby improving its balance sheet

William Clay Ford Jr.

(资产负债表). This arrangement took effect on January 1, 2010. As a sign of its currently strong cash position, Ford contributed its entire current liability (estimated at approximately US $5. 5 billion as of December 31, 2009) to the VEBA in cash, and also pre-paid US $500 million of its future liabilities to the fund. The agreement also gives hourly workers the job security they were seeking by having the company commit to substantial investments in most of its factories.

The automaker reported the largest annual loss in company history in 2006 of $12. 7 billion, and estimated that it would not return to profitability until 2009. However, Ford surprised Wall Street in the second quarter of 2007 by posting a $750 million profit. Despite the gains, the company finished the year with a $2. 7 billion loss, largely attributed to finance restructuring at Volvo.

On June 2, 2008, Ford sold its Jaguar and Land Rover operations to Tata Motors for $2. 3 billion.

In January 2009, Ford reported a $14. 6 billion loss in the preceding year, a record for the company. The company retained sufficient liquidity to fund its operations. Through April 2009, Ford's strategy of debt for equity exchanges(股票交易所) erased $9. 9 billion in liabilities (负债) (28% of its total) in order to leverage its cash position. These actions yielded Ford a $2. 7 billion profit in fiscal year 2009, the company's first full-year profit in four years.

In 2012, Ford's corporate bonds were upgraded from junk to investment grade again, citing sustainable, lasting improvements.

In February 2017, Ford Motor Co. acquired majority ownership of Argo AI (阿尔戈人工智

能）, an artificial-intelligence startup.

At the beginning of 2018, Jim Hackett was announced to replace Mark Fields as CEO of Ford Motor. Mr. Hackett most recently oversaw the formation of Ford Smart Mobility, a unit responsible for experimenting with car-sharing programs, self-driving ventures and other programs aimed at helping the 114-year-old auto maker better compete with Uber Technologies Inc., Alphabet Inc. and other tech giants looking to edge in on the auto industry.

On April 25, 2018, Ford announced that it will discontinue passenger cars in the North American market in the next four years, except for the Mustang and the Focus Active, due to declining demand and profitability. But on August 31, 2018, Ford announced that the Focus Active will not go on sale in North America because of the tariffs that would be placed on vehicles built overseas, as the Focus Active is being built in China and later announced in a statement on September 10, 2018 that they have no plans to build it in the United States.

(*If you want to find more information about this corporation, please log on https://en. wikipedia. org/wiki/Ford_Motor_Company.*)

1. *Read the passage quickly by using the skills of skimming and scanning, and choose the best answer to the following questions.*

    1) Under which brand does Ford Motor Company sell its most luxury cars? _____

        A. Lincoln   B. Cadillac

        C. Volvo   D. Mercury

    2) Which of the following statements is NOT true? _____

        A. Ford Motor Company acquired Jaguar and Land Rover in 1989 and 2000 respectively.

        B. The Cadillac Motor Company used to be a subsidiary of Ford Motor Company.

        C. Ford owned the Swedish automaker Volvo from 1999 to 2010.

        D. Ford discontinued the Mercury brand in 2011.

    3) Henry Ford's first attempt at a car company was _____.

        A. the Ford Motor Company in 1903

        B. the Ford Motor Company in 1901

        C. the Henry Ford Company in 1901

        D. the Cadillac Company in 1902

    4) Ford launched its first low-priced car with _____.

        A. a gasoline-powered engine in 1903

        B. mass-produced Model T in 1908

        C. safety glass in the windshield in 1927

        D. a V8 engine in 1932

    5) Some innovations were included in the lifeguard safety package offered by Ford in 1956. Which of the following is NOT one of the innovations? _____

        A. a standard deep-dish steering wheel

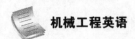
B. rear seatbelts

C. child-proof door locks

D. an optional padded dash

6) On January 23, 2006, Ford unveiled its "The Way Forward" plan, which floolwing was not included._____

   A. introducing a range of new vehicles

   B. resizing the company to match market realities

   C. discontinuing some unprofitable and inefficient models

   D. shutting down 14 factories and cutting a number of jobs

7) At the beginning of the 21st century, Ford announced that it will cooperate with _____ to survey the future of plug-in hybrids.

   A. Toyota   B. Southern California Edison

   C. The National Grid   D. GM

8) At the beginning of the 21st century, Ford sustained the largest annual loss in _____.

   A. 2006   B. 2007

   C. 2008   D. 2009

9) In _____, Ford's corporate bonds were upgraded from junk to investment grade again.

   A. 2007   B. 2009

   C. 2010   D. 2012

10) The Ford Smart Mobility aimed to _____.

    A. experiment with car-sharing program

    B. start self-driving ventures

    C. enhance its earnings

    D. help Ford better compete with the tech giants

2. *In this part, the students are required to make oral presentations on either of the following topics.*

   1) measures taken by Ford to return to profitability

   2) lessons from Ford's ups and downs

习题答案

Unit Seven　3D Printing

I. Pre-class Activity

Directions: *Please read the general introduction about* **Chuck Hull** *and tell something more about the great scientist to your classmates.*

Chuck Hull (Charles W. Hull is the co-founder(联合创始人), executive vice president (执 行 副 总 裁) and chief technology officer (首席技术官) of 3D Systems. He is the inventor of the solid imaging process(固体成像过程) known as stereolithography （ 立 体 光 刻 ） (3D Printing ） , the first commercial rapid prototyping(成型) technology, and the STL file format (文件格式). He is named on more than 60 U. S. patents(专利) as well as other patents around the world in the fields of ion optics(离子光学) and rapid prototyping. He was inducted into the National Inventors Hall of Fame(名人堂) in 2014 and in 2017 was one of the first inductees into the TCT Hall of Fame.

Chuck Hull was born on May 12, 1939 in Clifton, Colorado, the son of Lester and Esther Hull. His early life was spent in Clifton and Gateway, Colorado. He graduated from Central High School in Grand Junction, Colorado. Chuck received a BS in engineering physics from the University of Colorado in 1961.

In 1986, commercial rapid prototyping was started by Chuck Hull when he founded 3D Systems in Valencia, California. Chuck Hull realized that his concept was not limited to liquids and therefore gave it the generic name "stereolithography" (3D printing), and filed broadpatent claims covering any "material capable of solidification" (能够凝固的材料) or "material capable of altering its physical state"(能够改变物理状态的材料).

The salary for his role as 3D Systems CTO was \$307,500 in 2011.

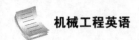

II. Specialized Terms

Directions: *Please memorize the following specialized terms before the class so that you will be able to better cope with the coming tasks.*

aesthetic n. 美的,美学

airbus n. 空中巴士

anthropometrics n. 人体测量学

assumed adj. 假定的;假装的

Atlantic adj. 大西洋的

atom n. 原子

attachment n. 附件;依恋;连接物;扣押
 财产

attachment point 附着点;接合点;起赔
 点;附加点

attributes v. 认为……属于

automated adj. 自动化的;机械化的

automatic identification 自动识别

automation n. 自动化技术,自动操作

back up v. 支持,援助;(资料)备份;倒
 退;裱

bandwidth n.[电子][物]带宽;[通信]频
 带宽度

based v. 立(基)于;以……为基础(base
 的过去式和过去分词)

battery n.[电]电池,蓄电池 n.[法]殴打 n.
 [军]炮台,炮位

biomechanics n. 生物力学;生物机械学

bit n. 少量;字节;一点 vt. 控制

blackmail n. 勒索,敲诈;勒索所得之款

bladders n. 膀胱

bluetooth headset 蓝牙耳机

boundary n. 边界;范围;分界线

bounds n. 界限

branch vt. 分支;出现分歧 vi. 分支;出现
 分歧

broadband n. 宽频;宽波段 adj. 宽频带的;

宽波段的;宽频通讯的

broadest adj. 宽广的

bronchial adj. 支气管的

built-in control system 内置控制系统

bulk n. 体积,容量;大多数,大部分;大块
 vt. 使扩大,使形成大量;使显得重要

bulk email 批量电子邮件;垃圾邮件

bureau n. 局,处;衣柜;办公桌

calculation n. 计算;估计;计算的结果;深
 思熟虑

cam n. 凸轮

capacity n. 能力;容量;资格,地位;生产力

carbide n. 碳化物

carbon n. 碳

carbon reduction 碳还原(法)

category n. 种类;分类

cells n. 细胞

coatings n.涂料

consciousness n. 意识;知觉

customization n. 定制

dense adj. 稠密的;浓厚的

jumbo adj. 巨大的

disruptive adj. 破坏的

fabricate vt. 制造

footprint n. 足迹;脚印

geometry n. 几何学

handicap n. 障碍;不利条件

hoover n. 吸尘器

implant n.[医]植入物

incorporate vt. 包含,吸收;体现

injection n. 注射

inspiring adj. 鼓舞人心的

interface n. 界面

intricate adj. 复杂的

jet n. 喷射,喷嘴;喷气式飞机

kidney n.[解剖]肾脏

layering v. 分层而成

literally adv. 按字面上

manipulate vt. 操纵;操作

mega adj. 许多的;宏大的

mimic adj. 模仿的,类似的,拟态的

molded adj. 模塑的;造形的

organism n. 有机体;生物体;微生物

prosthetic adj. 假体的

selectric n. 电动打字机

self-regulate 自我调节

set n. 集合,设置

sharp adj. 急剧的;锋利的;强烈的;敏捷的;刺耳的

signal n. 信号;暗号;导火线

skeleton n. 骨架,骨骼

software n. 软件

solenoid n. 螺线管

specific adj. 特定的

spin vi. 旋转;纺纱;吐丝;晕眩

spinning n. 纺纱

spray n. 喷雾;喷雾器;水沫; vt. 喷射;

vi. 喷

sub-discipline n. 学科的分支;副学科

synthetic adj. 综合的;合成的,人造的

system n. 制度;体制;系统;方法

task n. 任务;工作 vt. 分派任务

technician n. 技术员;工作人员

term n. 术语;学期;期限;条款

test chamber 试验箱;试验室

test cock 试验旋塞

test condition 试验条件

testing apparatus 测试装置

testing bench 试验台

test microphone 测试传声器

test specimen tube 试样筒(管)

test stand 试验台

tetrahedral angle 四面角

theory n. 原理;理论

thermodynamic adj. 热力学的;使用热动力的

tool n. 工具

valve n.阀;[解剖]瓣膜

variant adj. 不同的;多样的

vision n. 美景;眼力;幻象

weird adj. 怪异的;不可思议的

III. Watching and Listening

Task One   A 3D Printing Jumbo Jet

New Words

jumbo adj. 巨大的

jet n. 喷射,喷嘴;喷气式飞机

vision n. 美景;眼力;幻象

disruptive adj. 破坏的

inspiring adj. 鼓舞人心的

Atlantic adj. 大西洋的

airbus n. 空中巴士

incorporate vt. 包含,吸收;体现

mimic adj. 模仿的,类似的,拟态的

skeleton n. 骨架,骨骼

weird adj. 怪异的;不可思议的

视频链接及文本

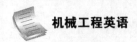

mega adj. 许多;宏大的

anthropometrics n. 人体测量学

dense n. 界面

consciousness n. 有机物

synthetic adj. 综合的;合成的,人造的

Exercises

1. *Watch the video for the first time and choose the best answer to the following questions.*

1) There are a lot of people who created their own _____ about the future.

   A. ideas
   B. idea
   C. visions
   D. vision

2) How long does it take to fly across the Atlantic Ocean? _____

   A. one day
   B. half day
   C. one and a half days
   D. two days

3) What kind of aircraft are we talking about here? _____

   A. Airbus A380
   B. Airbus A300
   C. Airbus A310
   D. Airbus A319

4) What are not the main customers of the future? _____

   A. old
   B. young
   C. men
   D. women

5) What value is not mentioned about a more sustainable future of aviation? _____

   A. social value
   B. religious value
   C. environmental value
   D. economic value

2. *Watch the video again and decide whether the following statements are true or false.*

1) We can predict the future. We only can create a vision of the future. (   )

2) So our children are getting larger, but at the same time we are growing into different directions. (   )

3) So we see a clear need of active health promotion, especially in the case of the young people. (   )

4) At the one hand, we have 1.2 kilos, and at the other hand 0.6 kilos. (   )

5) So we have very different seats which adapt to the shape of the future passenger, with the different anthropometrics. (   )

3. *Watch the video for the third time and fill in the following blanks.*

What do we know about the _____? Difficult question, simple answer: _____. We cannot _____ the future. We only can create a _____ of the future, how it might be, a vision which _____ disruptive ideas, which is inspiring, and this is the most important reason which breaks the chains of _____ thinking. There are a lot of people who created their own vision about the future, for instance, this vision here from the _____ century.

It says here that this is the _____ plane of the future. It takes only one and a half days to cross the _____ Ocean. Today, we know that this future vision didn't come true. So this is our largest airplane which we have, the Airbus A380, and it's quite huge, so a lot of people fit in there and it's _____ completely different than the vision I've shown to you.

4. *Share your opinions with your partners on the following topics for discussion.*

　　1) Do you like the lecture about a 3D printing jumbo jet? Why do you enjoy such a video? Please summarize the features of a 3D printing jumbo jet.

　　2) Can you use a few lines to list what's your understanding about 3D printing? Please use an example to clarify your thoughts.

Task Two　Introduction and Application of 3D Printing

New Words

視頻鏈接及文本

hoover n. 吸尘器

geometry n. 几何学

literally adv. 按字面上

fabricate vt. 制造

molded adj. 模塑的;造形的

en masse adv.(法)全体地;一同地

injection n. 注射

customization n. 定制

variant adj. 不同的;多样的

manipulate vt. 操纵

attribute v. 认为……属于

bound n. 界限

carbon n. 碳

footprint n. 足迹;脚印

aesthetic n. 美的,美学

prosthetic adj. 假体的

handicap n. 障碍;不利条件

coatings n.涂料

implant n.[医]植入物

bronchial adj. 支气管的

intricate adj. 复杂的

layering v. 分层

cell n. 细胞

bladder n. 膀胱

kidney n.[解剖]肾脏

valve n.阀;[解剖]瓣膜

Exercises

1. *Watch the video for the first time and choose the best answer to the following questions.*

　　1) Which of the following is true about RepRap machine? _____

　　　　A. 3D printer　　　　　　　　B. 2D printer

　　　　C. desktop printer　　　　　　D. printer

　　2) There's no need to do a run of thousands of millions or send that product to be injection molded in _____.

　　　　A. India　　　　　　　　　　B. China

　　　　C. Thailand　　　　　　　　　D. Japan

　　3) Which brands are not mentioned in the article? _____

　　　　A. Prada　　　　　　　　　　B. Smart car

　　　　C. Nike　　　　　　　　　　 D. Adidas

　　4) Which of the following is true about design your lamp? _____

　　　　A. price　　　　　　　　　　B. material

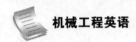
C. color   D. shape

    5) According to Dr. Anthony Atala, which one hasn't been working on layering cells to create body parts? _____

    A. bladders   B. valves

    C. head   D. kidneys

2. *Watch the video again and decide whether the following statements are true or false.*

    1) 3D printing can be also used to download spare parts from the web. (　　)

    2) What you could do is really influence your product now and shape-manipulate your product. (　　)

    3) The piece of software will keep an individual out of the bounds of the possible. (　　)

    4) We can create very specific prosthetics for that individual. (　　)

    5) Technology is going to cause a manufacturing revolution. (　　)

3. *Watch the video for the third time and write at least three products and their application field of 3D printing.*

Products	Field
Hoover	Home Application

4. *Share your opinions with your partners on the following topics for discussion.*

    1) What's your impression of 3D printing?

    2) Make a discussion about the operation and application of 3D printing in the manufacturing.

IV. Talking

Task One　Classical Sentences

Directions: *In this section, some popular sentences are supplied for you to read and to memorize. Then, you are required to simulate and produce your own sentences with reference to the structure.*

General Sentences

    1. Would you please tell John that I'm here?

    你能告诉约翰我在这吗?

    2. Would you help me lift this heavy box?

    你能帮我将这个重盒子抬起来吗?

3. Please ask John to turn on the lights.

请让约翰把灯打开。

4. Get me a hammer from the kitchen, will you?

从厨房里给我拿个锤子, 好吗？

5. Would you mind mailing this letter for me?

你愿意帮我发这封邮件吗？

6. If you have time, will you call me tomorrow?

如果你有时间, 明天给我打电话好吗？

7. Please pick up those cups and saucers.

请将那些杯子和碟子收拾好。

8. Will you do me a favor?

你能帮我个忙吗？

9. Excuse me, sir. Can you give me some information?

先生, 打扰一下, 你能告诉我一些信息吗？

10. Do you happen to know Mr. Cooper's telephone number?

你知道库珀先生的电话号码吗？

11. Would you mind giving me a push? My car has stalled.

你能帮我推一下车吗？ 我的车抛锚了。

12. Would you be so kind as to open this window for me? It's stuffy.

你能帮我把窗户打开吗？ 好闷人。

13. If there's anything else I can do, please let me know.

如果还有我能做的事情, 请告诉我。

14. This is the last time I'll ever ask you to do anything for me.

这是最后一次麻烦你为我办事了。

15. I certainly didn't intend to cause you so much inconvenience.

我真不想给你带来这么多不便。

16. Would you please hold the door open for me?

帮我开门好吗？

17. You're very kind to take the trouble to help me.

你真是太好了, 不嫌麻烦来帮我。

18. I wish I could repay you somehow for your kindness.

但愿我能以某种方式报答你的好意。

19. I'm afraid it was a bother for you to do this.

做这件事恐怕会给你带来很多麻烦。

20. He'll always be indebted to you for what you've done.

对你所做的事情, 他总是感激不尽。

21. Could you lend me ten dollars? I left my wallet at home.

能借我十美元吗？ 我把钱包忘在家里了。

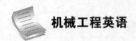

22. I'd appreciate it if you would turn out the lights. I'm sleepy.

如果你能将灯关掉的话,我会很感激。我好困。

23. You're wanted on the telephone.

有你的电话。

24. What number should I dial to get the operator?

我想接通接线员的电话,我应该拨打什么号码?

25. The telephone is ringing, would you answer it, please?

电话铃响了,你能接下电话吗?

26. Would you like to leave a message?

你想留下什么口信儿吗?

27. I have to hang up now.

我现在得挂电话了。

28. Put the receiver closer to your mouth. I can't hear you.

将话筒靠近你的嘴巴,我听不见你的声音。

29. Would you mind calling back sometime tomorrow?

你介意明天再打过来吗?

30. I almost forgot to have the phone disconnected.

我差点忘记挂断电话了。

31. It wasn't any bother. I was glad to do it.

一点儿都不麻烦,我很乐意做这件事。

32. There's just one last favor I need to ask of you.

还有最后一件事需要你的帮助。

33. I'd be happy to help you in any way I can.

很高兴我能尽我所能地帮助你。

34. Please excuse me for a little while. I want to do something.

对不起,稍等会儿,我有点事要办。

35. I didn't realize the time had passed so quickly.

我没有意识到时间过得这样快。

36. I've got a lot of things to do before I can leave.

走之前,我有很多事情要做。

37. For one thing, I've got to drop by the bank to get some money.

首先,我得去银行取些钱。

38. It'll take almost all my savings to buy the ticket.

买这张票几乎花掉了我所有积蓄。

39. Oh, I just remembered something! I have to apply for a passport.

我记起一件事,我得去申请个护照。

40. It's a good thing you reminded me to take my heavy coat.

你提醒我带件厚衣服,真是太好了。

41. I would never have thought of it if you hadn't mentioned it.
要是你不提到它,我几乎想不起来了。

42. I'll see you off at the airport.
我会去机场送你。

43. Let's go out to the airport. The plane landed ten minutes ago.
我们去机场吧。飞机在十分钟前就已经着陆了。

44. There was a big crowd and we had difficulty getting a taxi.
这里人很多,我们很难打到车。

45. They're calling your flight now. You barely have time to make it.
他们现在在广播你的班机起飞时间。你勉强来得及赶上。

46. You'd better run or you're going to be left behind.
你最好跑过去,不然就会被落下了。

47. Don't forget to call us to let us know you arrived safely.
你安全到达后,别忘记打电话报个平安。

48. I'm sure I've forgotten something, but it's too late now.
我确定我忘记了什么东西,但是现在太晚了。

49. Do you have anything to declare for customs?
你有什么需要报关的吗?

50. You don't have to pay any duty on personal belongings.
你的私人物品不用交税。

Specialized Sentences

1. Businesses are now trying to use new technology, such as 3D printing and augmented reality, to bring customers back into stores.
企业正在尝试使用 3D 打印和增强现实的新技术,吸引客户到商店购物。

2. Computer company Intel, is experimenting with 3D printing.
计算机公司英特尔正在试验 3D 打印技术。

3. The 3D printers print out clothing, made just for the individual in size and style, in about 45 minutes.
3D 打印机在大约 45 分钟内就能够打印出适合个人尺寸和款式的服装。

4. Mazdzer's success could have been the 3D printer technology, which his team used to make its equipment.
麦斯德赛成功的另一个原因可能是他的团队用来制造设备的 3D 打印机技术。

5. The term 3D is short for three dimensional—meaning an object with length, width and height.
术语 3D 是三维的缩写——具有长度、宽度和高度的物体。

6. In 3D printing, 3D models are first created as files, or documents, on a computer.
在 3D 打印中,首先在计算机上将三维模型创建为文件或文档。

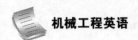

7. The printer then uses a substance like plastic or metal to create physical objects.

打印机使用诸如塑料或金属之类的物质来制作物体。

8. The process involves making one layer of material at a time until the objects reach full form.

该过程涉及一次制作一层材料,直到物体达到完整形状。

9. 3D printing technology was used to create tools for making molds in the shape of Olympians on top of a sled.

3D 打印技术被用于创建工具,用于在雪橇顶部制作奥运选手的模具。

10. Officials from USA Luge say 3D printing can greatly simplify the process, speeding up production of parts to within hours.

美国 Luge 公司的官员表示,3D 打印可以大大简化流程,将零件生产速度提高到数小时内。

11. 3D printing works well for making last-minute changes that require exactness.

3D 打印可以很好地在最后时刻进行所需的精确性修改。

12. Other sports represented at the 2008 Olympics have used 3D printing to improve their equipment.

2008 年奥运会期间的其他体育项目使用 3D 打印来改进他们的设备。

13. Snowboard maker CAPiTA used 3D printers to build stronger sidewalls for its snowboards.

滑雪板制造商 CAPiTA 使用 3D 打印机为其滑雪板构建更坚固的侧壁。

14. I just want to show you here what we are doing today.

我只想给大家展示我们正在研究的内容。

15. What do you think of 3D printing technology?

你觉得 3D 打印技术怎么样?

16. This here is an equal bracket for the same purpose.

这是一个用途相同的相似支架设计。

17. So this technology, 3D printing, and new design rules, really help us to reduce the weight.

所以 3D 打印和新的设计规则帮助我们减轻了重量。

18. So how does nature build its components and structures?

那么大自然是如何构建各个部分和结构的呢?

19. And the same approach can be applied to technology as well.

同样的方法也能运用到技术上。

20. Our building block is carbon nanotubes.

我们的构建基块是碳纳米管。

21. To create a large, rivet-less skeleton at the end of the day.

最终制造出一个巨型、无缝的构架。

22. How this looks in particular, you can show it here.

这看起来很特别,你可以在这里展现。

23. Imagine you have carbon nanotubes growing.
   想象一下在 3D 打印机里正在制作碳纳米管。

24. Carbon nanotubes are embedded inside a matrix of plastic.
   碳纳米管被嵌入一个塑料基体中。

25. You take this wood and make morphological optimization.
   你拿这块木头进行形态优化。

26. So you make structures, sub-structures which allows you to transmit electrical energy or data.
   然后制作架构以及次级架构,以便传输电能或数据。

27. And now we take this material, combine this with a top-down approach, and build bigger and bigger components.
   现在我们采用这种材料,结合自顶向下的方法,构建越来越大的组件。

28. And finally, this bionic structure, which is covered by a transparent biopolymer membrane.
   最后,这种仿生结构由透明生物聚合物膜覆盖。

29. You can download products from the web.
   你可以从网上下载产品。

30. There are many different technologies, used in the 3D printing process.
   3D 打印过程中使用了许多不同的技术。

31. We can actually build for you, very rapidly, a physical object.
   我们可以很快为你建造一个实物。

32. This is through an emerging technology called 3D printing.
   这是通过一种叫 3D 打印的新兴技术实现的。

33. This is a 3D printer.
   这是一台 3D 打印机。

34. You would take data, like the data of a pen.
   你可以拿数据,像这支笔的数据。

35. Product data can be presented geometrically in three dimensions.
   产品数据可以用三维几何展示。

36. We would pass that data with material into a machine.
   我们会把数据和材料一起输入机器。

37. The machine is programmed to print out the product layer by layer.
   机器按照程序一层层打印出来。

38. And we can take out the physical product.
   然后我们就造出了这个产品。

39. These data gets sent to a machine.
   这些数据被发送到一台机器。

40. Professional product designer create a product in 3D.
   专业的产品设计师制造了一个 3D 产品。

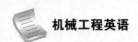

41. It will change and disrupt the landscape of manufacturing.

它将改变并打破制造业的格局。

42. The machine slices the data into two-dimensional representations of that product.

这台机器将数据分解为该产品的二维表示。

43. The machine typically reads CAD data.

这台机器通常读取 CAD 数据。

44. This material that's deposited either starts as a liquid form or a material powder form.

这种沉淀的物质要么是液态的,要么是粉状的。

45. The bonding process can happen by either melting and depositing or depositing then melting.

键合过程可以通过熔化、沉积或沉积后熔化来实现。

46. 3D printers are a new generation of machines that can make everyday things.

3D 打印机是新一代的机器,可以制造日常用品。

47. In this case, we can see a laser sintering machine developed by EOS.

在这个例子中,我们可以看到 EOS 造的一个激光烧结机。

48. A 3D printer can make pretty much anything from ceramic cups to plastic toys, metal machine parts, stoneware vases, fancy chocolate cakes or even human body parts.

3D 打印机几乎可以制造任何东西,从陶瓷杯到塑料玩具、金属机械零件、石器花瓶、精美的巧克力蛋糕,甚至人体部件。

49. This course introduces the theory and technology of micro/ nano fabrication.

本课程介绍了微米/纳米制造理论和技术。

50. The main reason you should do this is for self progression.

你应该这样做的主要原因是为了自我进步。

Task Two   Sample Dialogue

Directions: *In this dialogue, you are going to read several times the following sample dialogue about the relevant topic. Please pay special attention to five Cs (culture, context, coherence, cohesion and critique) in the dialogue and get ready for a smooth communication in the coming task.*

Job Interview

(*Mr. Bao is going to attend an interview with Mr. Liu.*)

Mr. Liu: Come in , please.

Mr. Bao: Good afternoon, Mr. Liu.

Mr. Liu: Good afternoon. Have a seat, please.

Mr. Bao: Thank you very much.

Mr. Liu: Are you Mr. Bao?

Mr. Bao: Yes, I am.

Mr. Liu: I have read your resume. I know you have worked for 3 years. Why did you choose to major in mechanical engineering?

Mr. Bao: Many factors lead me to majoring in mechanical engineering. The most important factor is that I like tinkering with machines.

Mr. Liu: What are you interested in about mechanical engineering?

Mr. Bao: I like designing products and one of my designs received an award. Moreover, I am familiar with 3D printing.

Mr. Liu: Great. Then what is your technical post title now?

Mr. Bao: I'm a senior mechanical design engineer.

Mr. Liu: Do you take the original certificate with you?

Mr. Bao: Yes. Here it is.

Mr. Liu: Why did you decide to apply for this position?

Mr. Bao: Your company has a very good reputation. I'm very interested in the field of your company.

Mr. Liu: Well, thanks. I'll let you know the result of the interview as soon as possible. Goodbye.

Mr. Bao: Thank you. I do hope the answer will be favorable. Goodbye.

Task Three Simulation and Reproduction

Directions: *The class will be divided into three major groups, each of which will be assigned a topic. In each group, some students may be the teacher, while others may be students. In the process of discussion, please observe the principles of cooperation, politeness and choice of words. One of the groups will be chosen to demonstrate the discussion to the class.*

1) 3D printing in our daily life
2) What do you think of 3D printing technology?
3) the importance of learning 3D printing

Task Four Discussion and Debate

Directions: *The class will be divided into two groups. Please choose your stand in regard to the following controversy and support your opinions with scientific evidences. Please refer to the specialized terms and classical sentences in the previous parts of this unit.*

3D printing are particularly useful in a wide variety of applications, such as medical treatment, aerospace, car industry. Does technology is going to cause a manufacturing revolution? Why? Why not?

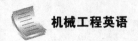

V. After-class Exercises

1. *Match the English words in Column A with the Chinese meaning in Column B.*

A	B
1) fabricate	A) 模仿的,类似的
2) molded	B) 几何学
3) manipulate	C) 吸尘器
4) carbon	D) [医] 植入物
5) footprint	E) [涂料] 涂料
6) coatings	F) 足迹;脚印
7) implant	G) 碳
8) hoover	H) 操纵;操作
9) geometry	I) 模塑的;造形的
10) mimic	J) 制造

2. *Fill in the following blanks with the words or phrases in the word bank, a change the forms if it's necessary.*

substance	files	customers	equipment	size
stronger	simplify	3D	create	height

1) Snowboard maker CAPiTA used 3D printers to build _____ sidewalls for its snowboards.

2) Businesses are now trying to use new technology, such as 3D printing and augmented reality, to bring _____ back into stores.

3) Other sports represented at the 2008 Olympics have used 3D printing to improve their _____.

4) The 3D printers print out clothing, made just for the individual in _____ and style, in about 45 minutes.

5) Officials from USA Luge say 3D printing can greatly _____ the process, speeding up production of parts to within hours.

6) Mazdzer's success could have been the _____ printer technology, which his team used to make its equipment.

7) 3D printing technology was used to _____ tools for making molds in the shape of Olympians on top of a sled.

8) The term 3D is short for three dimensional—meaning an object with length, width and _____.

9) In 3D printing, 3D models are first created as _____, or documents, on a computer.

10) The printer then uses as _____ like plastic or metal to create physical objects.

3. *The following is the main structure of 3D printer. Can you write the corresponding parts?*

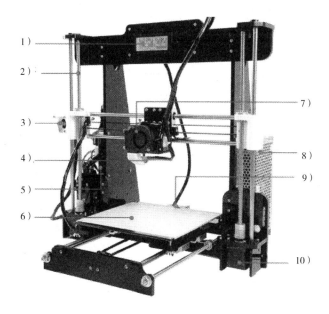

1) _____ 2) _____ 3) _____ 4) _____ 5) _____
6) _____ 7) _____ 8) _____ 9) _____ 10) _____

4. *Translate the following sentences into English.*

1）我们可以实现最新的人机互动。

_____

_____

2）我们想要将这些想法融入产品中。

_____

_____

3）因此我们将这些需要与技术主题相结合。

_____

_____

4）我们如何能创造更多的光照？

_____

_____

5）计算机公司英特尔正在试验 3D 打印。

_____

_____

5. *Please write an essay of about* 120 *words on the topic：**Application of 3D printing in our life.** Some specific examples will be highly appreciated and you have to watch out the spelling of some specialized terms you have learnt in this unit.*

_____

_____

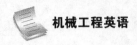

_____

_____

_____

_____

_____

_____

_____

_____

VI. Additional Reading

The History and Organizational Structure of Stratasys

Stratasys, Ltd. is a manufacturer（制造商）of 3D printers and 3D production systems for office-based rapid prototyping（原型制作）and direct digital manufacturing solutions. Engineers use Stratasys systems to model complex geometries（几何形状）in a wide range of thermoplastic（热塑性）materials, including: ABS, polyphenyl sulfone（聚苯砜,PPSF）, polycarbonate（聚碳酸酯,PC）and ULTEM 9085.

Stratasys manufactures in-office prototyping and direct digital manufacturing systems for automotive, aerospace, industrial, recreational（娱乐）, electronic, medical and consumer product OEMs.

History

Stratasys was founded in 1989, by S. Scott Crump and his wife Lisa Crump in Eden Prairie（伊甸草原）, Minnesota（明尼苏达州）. The idea for the technology came to Crump in 1988 when he decided to make a toy frog for his young daughter using a glue gun loaded with a mixture of polyethylene（聚乙烯）and candle wax. He thought of creating the shape layer by layer and of a way to automate the process. In April 1992, Stratasys sold its first product, the 3D Modeler.

In October 1994, Stratasys had an initial public offering on NASDAQ（募股）; the company sold 1. 38 million shares of common stock at $5 per share, netting approximately $5. 7 million.

In January 1995, Stratasys purchased IBM's rapid prototyping intellectual property and other assets, and employed 16 former（前）IBM engineers, who had been developing a small 3D printer that relied on an extrusion system very similar to Crump's patented fused deposition modeling(熔融沉积建模,FDM) technology.

In 2003, Stratasys fused deposition modeling (FDM) was the best-selling rapid prototyping technology. FDM is a process that the company patented(专利的), which is used to produce three-dimensional(空间的) parts directly from 3D CAD files layer-by-layer(逐层) for use in design verification(验证), prototyping(原型设计), development and manufacturing.

In 2007, Stratasys supplied 44% of all additive fabrication(添加剂制造) systems installed worldwide, making it the unit market leader for the sixth consecutive(连贯的) year.

In January 2010, Stratasys signed an agreement with HP to manufacture HP-branded 3D printers. In August 2012, the HP manufacturing and distribution agreement was discontinued.

In May 2011, Stratasys announced that they had acquired Solidscape, a leader in high-precision(精度) 3D printers for lost wax casting(失蜡铸造) applications.

In 2012, an unrelated project to produce a working firearm(火枪) by 3D printing was intended to use a Stratasys printer. Stratasys refused to permit this and withdrew(撤回) its license(许可) for use of the printer, citing that it did not allow its printers "to be used for illegal purposes".

In 2014, the Israeli(以色列) fashion designer Noa Raviv featured grid pattern centered couture garments which were created employing Stratasys' 3D printing technology. Some selections from the aforementioned collection were exhibited in 2016 at the exhibition "Manus X Machina" at the Anna Wintour Costume Center at New York City's Metropolitan Museum of Art.

Picture of Stratasys Corporate Headquarters

Merger(合并) with Objet

On April 16, 2012, Stratasys announced that it agreed to merge with privately held Objet Ltd., a leading manufacturer of 3D printers based in Rehovot, Israel, in an all-stock

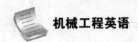

transaction. Stratasys shareholders were expected to own 55 percent of the combined company, and Objet shareholders would own 45 percent. The merger was completed on December 3, 2012; the market capitalization of the new company was approximately $3. 0 billion.

Acquisition(收购) of MakerBot, Solid Concepts and Harvest Technologies

On June 19, 2013, MakerBot Industries announced that it was purchased by Stratasys for $403 million.

On April 2, 2014, Stratasys announced that they had entered into definitive agreements to acquire Solid Concepts and Harvest Technologies, which will be combined with RedEye, its existing digital manufacturing service business, to establish a single additive manufacturing services business unit. The acquisition was finalized on July 15, 2014.

Investment in Massivit

On February, 2016, Stratasys announced an investment in Israeli's company Massivit 3D Printing Technologies, to promote and deploy(部署) Massivit's proprietary super-sized 3D printing solutions.

3D Car Production Systems

In 2014, Stratasys prototyped an electric car with fully 3D-printed exterior panels(外部面板), and a few printed interior parts. Development took one year, and parts(零件) were constructed using a Stratasys Objet 1000.

Urbee

Urbee is the name of the first car in the world car mounted using additive(增材) manufacturing technology (its bodywork and its car windows were "printed"). Created in 2010 through the partnership with the Canadian engineering group KOR Ecologic, it is a hybrid vehicle with futuristic look.

(*If you want to find more information about this corporation, please log on https://en. wikipedia. org/wiki/Stratasys.*)

1. *Read the passage quickly by using the skills of skimming and scanning, and choose the best answer to the following questions.*

   1) Stratasys was founded in _____.

      A. 1989   B. 1988

      C. 1992   D. 1994

   2) What's the first product of Stratasys? _____

      A. printer   B. 3D printer

      C. 3D Modeler   D. Modeler

   3) In _____, Stratasys fused deposition modeling (FDM) was the best-selling rapid prototyping technology.

A. 2002   B. 2003

C. 2004   D. 2005

4) In 2014, which country created 3D printing technology? _____

   A. America   B. British

   C. Israeli   D. Canada

5) Which city holds the exhibition "Manus X Machina" at the Anna Wintour Costume
   Center? _____

   A. Beijing   B. London

   C. New York   D. Japan

6) When does Stratasys merger with Objet? _____

   A. June 19, 2013   B. April 16, 2012

   C. April 2, 2014   D. February, 2016

7) How many shares do the shareholders of Stratasys expect to own? _____

   A. 55 percent   B. 45 percent

   C. 50 percent   D. 40 percent

8) On June 19, 2013, MakerBot Industries announced that it was purchased by Stratasys
   for _____.

   A. $400 million   B. $403 million

   C. $430 million   D. $413 million

9) In 2014, Stratasys prototyped an electric _____ with fully 3D-printed exterior panels.

   A. printer   B. parts

   C. objects   D. car

10) _____ is the name of the first car in the world car mounted using additive
    manufacturing technology.

    A. Urbee   B. Tesla

    C. Audi   D. BMW

2. *In this part, the students are required to make an oral presentation on either of the following
   topics.*

   1) the history of Stratasys

   2) the lessons from Stratasys' development history

习题答案

Unit Eight Industrial Robots

I. Pre-class Activity

Directions: *Please read the general introduction about **George Devol** and tell something more about the great scientist to your classmates.*

George Devol applied for the first robotics patents in 1954 [granted(授予)in 1961]. The first company to produce a robot was Unimation, founded by Devol and Joseph F. Engelberger in 1956. Unimation robots were also called programmable transfer machines(可编程传输机器)since their main use at first was to transfer objects from one point to another, less than a dozen feet or so apart. They used

hydraulic actuators(液压执行器) and were programmed in joint coordinates(关节坐标), i. e. the angles of the various joints were stored during a teaching phase and replayed in operation. They were accurate to within 1/10,000 of an inch (note：although accuracy is not an appropriate measure for robots, usually evaluated in terms of repeatability—see later). Unimation later licensed their technology to Kawasaki Heavy Industries and GKN, manufacturing Unimates in Japan and England respectively. For some time Unimation's only competitor was Cincinnati Milacron Inc. of Ohio.

In 1969 Victor Scheinman at Stanford University invented the Stanford arm, an all-electric (全电动), 6-axis(轴) articulated robot designed to permit an arm solution. This allowed it accurately to follow arbitrary paths in space and widened the potential use of the robot to more sophisticated applications such as assembly(装配) and welding(焊接). Scheinman then designed a second arm for the MIT AI Lab, called the "MIT arm".

II. Specialized Terms

Directions：*Please memorize the following specialized terms before the class so that you will be able to better cope with the coming tasks.*

actuator n.［电］制动器

anthropomorphize vt. 赋予人性

artificial adj. 人造的,人工的

assembly n. 装配;集会,集合

assembly robot 装配机器人

automatic guided vehicle 自动引导车

autonomous adj. 自主的

auxiliary adj. 辅助的,附加的

axis n. 轴

buffering n. 缓冲

centimeter n.［计量］厘米

chain n. 链条

civilization n. 文明;文化

closed-loop 闭环的

code n. 代码,密码

code conversion 代码转换

conscious adj. 有意识的

data n. 数据

digital controlled 数控的

hydraulic adj. 液压的;水力的;水力学的

droid n. 机器人

electronics n. 电子学

endpoint n. 末端

entertain vt. 娱乐

entrepreneurship n. 企业家精神

equipment n. 设备

explosive adj. 易爆的 n. 炸药

fiction n. 小说;虚构

fire robot 消防机器人

food industry robots 食品工业机器人

gear n. 齿轮

given program 给定程序

gripper n. 夹子,钳子;抓器,抓爪

handling robot 搬运机器人

heat dissipation 散热,热损耗

humanity n. 人类

ignore vt. 驳回诉讼;忽视

inaction n. 不活动;迟钝

incredibly adv. 难以置信地

indicate vt. 表明;指出

indicated value 指示值

indicating device 指示装置

indicating instrument 指示仪器

indicator diagram 示功图

industrial mechanics 工程力学

industrial process measurement and control
   instrument 工业测量与控制仪表

industrial robot 工业机器人

industrial unit 工业设备

inflammable adj. 易燃的,易怒的,易激
   动的

infrastructure n. 基础设施

intellectual adj. 智力的;聪明的;理智的

intelligent adj. 智能的

illusion n. 幻觉,错觉

interact v. 相互作用

interrupt v. 中断

joy-stick 操作杆

labour n. 劳动力,人工

larger-sized 大型的

lathe v. 用车床加工 n. 车床,机床

leitmotif adj. 不合理的

machining v.机械加工

mainframe n. 主机,主计算机

manipulator n. 机械手,操纵臂

marvel vt. 对……感到惊异

material handling 物料输送、原材料处理

mechanical adj. 机械的;力学的

mechanical body 机械本体

mental adj. 精神的;脑力的;疯的

mining robot 采矿机器人

mount v. 安装

open-loop 开环的

operational adj. 可操作的;运作的

painting robot 喷漆机器人

path n. 路,路径

physical adj.物理的;物质的

pioneer n. 先锋;拓荒者

plague n. 瘟疫;灾祸

pre-programmed adj. 预先编排程序的

presence n. 存在

priority n. 优先级

productive adj. 能生产的

programmable adj. 可计算机控制的,可程

序控制的

proximity n. 接近

radioactive adj. 放射性的,有辐射的

remote control 遥控器

rescue robot 救援机器人

revolutionize vt. 彻底改革,发动革命

ridiculously adv. 可笑地

robot dynamics 机器人动力学

sonar n. 声呐;声波定位仪

spot welding 点焊

spray v. 喷,喷洒

spray painting 喷漆

synchronization n. 同步化

transition n. 过渡;转变

vice n. 恶习;缺点

truck n. 卡车

welding adj. 焊接的 n. 焊接 v. 焊接;锻接

welding head 焊接头

welding robot 焊接机器人

workmanship n.手艺,工艺;技巧

III. Watching and Listening

Task One　Are the Robots Taking our Jobs?

New Words

autonomous adj. 自治的,自主的,自发的

plague n. 瘟疫,麻烦事,天灾

humanoid adj. 像人的

mockery n. 嘲弄,笑柄

vice n. 恶习,缺点 adj. 副的,代替的

darn v. 缝补

vouch v. 担保,证明

flex v. 折曲,弯曲

trajectory n. 轨道

视频链接及文本

Exercises

1. *Watch the video for the first time and choose the best answer to the following questions.*

　1) How many people drive trucks for a living in the United States? _____

　　　A. three million　　　　　　　　B. half million

　　　C. three and a half million　　　D. two million

　2) We can stimulate job growth by encouraging _____.

A. scholarship   B. leadership

C. partnership   D. entrepreneurship

3) Work saves us from three great evils, which of the following is not included? _____

A. freedom   B. boredom

C. vice   D. need

4) What has been the most important developments in human history from the speaker's point of view? _____

A. technology   B. system of philosophy

C. religiou   D. empire

5) During Industrial Revolution, what changed the world and influenced human history so much? _____

A. steam engine   B. train

C. car   D. bike

2. *Watch the video again and decide whether the following statements are true or false.*

1) Digital technologies are just impacting knowledge work. (   )

2) The speaker had the chance a little while back to ride in the Google autonomous bike, which is not as cool as he expected. (   )

3) There are varied answers in respond to the question that what have been the most important developments in human history. (   )

4) The speaker is pessimistic about the digital technologies which are going to take us into a dystopian future. (   )

5) The robots today still very good at fixing bridges. (   )

3. *Watch the video for the third time and fill in the following blanks.*

It's a wonderful question to ask and to start an endless _____ about, because some people are going to bring up systems of _____ in both the West and the _____ that have changed how a lot of people think about the world. And then other people will say, " No, _____, the big stories, the big developments are the _____ of the world's major _____, which have changed _____ and have changed and influenced how _____ people are living their lives. " And then some other folk will say, "Actually, what changes civilizations, what _____ them and what changes people's lives are _____, so the great developments in human history are stories of conquest and of war. " And then some cheery soul usually always pipes up and says, "Hey, don't forget about plagues. "

4. *Share your opinions with your partners on the following topics for discussion.*

1) Do you like the lecture from TED? Why do you enjoy such a lecture? Please summarize the features of TED.

2) Can you use a few lines to list what's your understanding about industrial robots? Please use an example to clarify your thoughts.

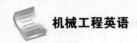

Task Two　The Best Robots Yet

New Words

leitmotif n. 主乐调,主题,主旨

anticipate v. 预料,期盼

pre-electronic adj. 前电子时代的

humanity n. 人类,人道

proximity n. 接近,邻近,亲近

conjurer n. 魔术师,巫师

intrigue n. 阴谋,诡计 v. 用诡计取得

devise v. 设计,发明

indistinguishable adj. 不能区别的

ethical adj. 道德的

spell n. 符咒,魔力 v. 拼写

fake adj. 假的,冒充的 n. 假货

unpredictability n. 不可预测性,不可预
　　见性

inaction n. 不活动,迟钝

warranty n. 保证,担保,(正当)理由

literally adv. 正确地,照字面地

视频链接及文本

Exercises

1. *Watch the video for the first time and choose the best answer to the following questions.*

　1) These mechanical performers were popular throughout _____.

　　　A. China　　　　　　　　B. Paris

　　　C. Italy　　　　　　　　D. Europe

　2) Who is EDI? _____

　　　A. an engineer　　　　　B. a robot

　　　C. a host　　　　　　　D. a robot's friend

　3) How tall is EDI? _____

　　　A. 176 centimeters　　　B. 360 centimeters

　　　C. 191 centimeters　　　D. 300 centimeters

　4) EDI became operational at TED in March _____.

　　　A. 2004　　　　　　　　B. 2040

　　　C. 2014　　　　　　　　D. 2010

　5) Which of the following is true according to the video? _____

　　　A. A robot may not be conscious of human's fragile frame and ignore our fears.

　　　B. A robot may harm humanity by inaction.

　　　C. A robot may not harm humanity, or by inaction allow humanity to come to harm.

　　　D. A robot may allow humanity to come to harm.

2. *Watch the video again and decide whether the following statements are true or false.*

　1) The perfect robot will be indistinguishable from the human. (　　)

　2) Robots can anticipate human actions. (　　)

　3) Humans and robots find it difficult to work in close proximity. (　　)

　4) Magic creates the illusion of an impossible reality. (　　)

　5) Human's intentions are predictable when they are irrational. (　　)

3. *Watch the video for the third time and fill in the following blanks of the table.*

Year/Number	Events
1921	
176	
300	
2014	

4. *Share your opinions with your partners on the following topics for discussion.*

1）What's your impression of a robot?

2）discussion of the operation and application of robots in the manufacturing

IV. Talking

Task One   Classical Sentences

Directions：*In this section, some popular sentences are supplied for you to read and to memorize. Then, you are required to simulate and produce your own sentences with reference to the structure.*

General Sentences

1. Children enter school at the age of five, don't they?
   孩子们五岁上学,是吗?

2. In elementary school, children learn to read and write.
   在小学,孩子们学习读和写。

3. In secondary school, children get more advanced knowledge.
   在中学,孩子们学到更多先进的知识。

4. In universities, students are trained to become teachers and engineers.
   在大学里,学生们被培养成老师和工程师。

5. He went to grade school in New York and high school in Chicago.
   他在纽约上小学,在芝加哥上中学。

6. In college I majored in science. What was your major?
   大学里我的专业是科学,你呢?

7. My sister graduated from high school. Graduation was last night.
   我姐姐中学毕业了。昨晚举行了毕业晚会。

8. I'm a graduate of Yale University. I have a Bachelor of Arts degree.
   我是一名耶鲁大学的毕业生。我获得了艺术学士学位。

9. If you expect to enter the university, you should apply now.

   如果你想上大学,现在就应该申请。

10. This is my first year of college. I'm a freshman.

    这是我大学的第一年,我是新生。

11. My uncle is a high school principal.

    我叔叔是一名中学校长。

12. What kind of grades did you make in college?

    你在大学里成绩怎么样?

13. During your first year of college, did you make straight As?

    你大学一年级时成绩全优吗?

14. My brother is a member of the faculty. He teaches economics.

    我的哥哥是一名老师,他教经济学。

15. John has many extracurricular activities. He's on the football team.

    约翰有许多课外活动。他参加了足球队。

16. I'm a federal employee. I work for the Department of Labor.

    我是一名联邦雇员。我在劳工部工作。

17. What kind of work do you do? Are you a salesman?

    你做什么工作? 是不是销售员?

18. As soon as I complete my training, I'm going to be a bank teller.

    一旦我完成了培训,我将成为一名银行出纳员。

19. John has built up his own business. He owns a hotel.

    约翰已经有了自己的生意,他拥有一家旅馆。

20. What do you want to be when you grow up?

    你长大后想做什么?

21. My son wants to be a policeman when he grows up.

    我儿子长大后想当警察。

22. I like painting, but I wouldn't want it to be my life's work.

    我喜欢绘画,但是我不会让绘画成为我终生的职业。

23. Have you ever thought about a career in the medical profession?

    你是否考虑过成为一名医药行业的从业人员?

24. My uncle was a pilot with the airlines. He has just retired.

    我的叔叔是航空公司的飞行员,他刚刚退休。

25. My brother's in the army. He was just promoted to the rank of major.

    我的哥哥在军队,他刚刚被提升为少校。

26. I have a good-paying job with excellent hours.

    我有一份工资高、工作时间理想的工作。

27. My sister worked as a secretary before she got married.

    我的姐姐结婚前是秘书。

28. George's father is an attorney. He has his own company.

乔治的父亲是律师,他拥有自己的公司。

29. He always takes pride in his work. He's very efficient.

他总是以他的工作为荣。他是个很能干的人。

30. Mr. Smith is a politician. He's running for election as a governor.

史密斯先生是一位政治家,他正在为竞选州长而奔波。

31. After a successful career in business, he was appointed ambassador.

他在生意中有了成就后,就被任命为企业代表。

32. Why is Mr. Smith so tired? Do you have any idea?

为什么史密斯先生这么累,你知道吗?

33. According to Mr. Green, this is a complicated problem.

听格林先生说,这是一个复杂的问题。

34. I wish you would give me a more detailed description of your trip.

我希望你能更详细地描述一下你的旅行。

35. We used to have a lot of fun when we were that age.

我们那么大时经常玩得很开心。

36. I never realized that someday I would be living in New York.

我从没有想过有一天我能住在纽约。

37. We never imagined that John would become a doctor.

我们从来没有想过约翰会成为一名医生。

38. I beg your pardon. Is this seat taken?

请问这个座位有人坐吗?

39. The waiter seems to be in a hurry to take our order.

服务员似乎急着要我们点东西。

40. I'd like my steak well-done.

我的牛排要全熟的。

41. —What kinds of vegetables do you have?

   —I'll have mashed potatoes.

   ——你想要什么蔬菜?

   ——我想要些土豆泥。

42. —What do you want?

   —I want a cup of coffee.

   ——你想要点什么?

   ——我想来杯咖啡。

43. —Which one would you like

   —this one or that one?

   ——你想要哪一个?

   ——这一个还是那一个?

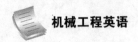

44. It doesn't matter to me.

随便都可以。

45. Would you please pass the salt?

麻烦把盐递过来好吗?

46. They serve good food in this restaurant.

这家餐馆的东西很好吃。

47. Are you ready for your dessert now?

现在可以吃点心了吗?

48. This knife/fork/spoon is dirty. Would you bring me a clean one, please?

这把刀/叉/勺脏了,麻烦你拿一个干净的来,好吗?

49. You have your choice of three flavors of ice cream. We have vanilla, chocolate, and strawberry.

你可以选择三种口味的冰淇淋,我们有香草味的、巧克力味的、草莓味的。

50. The restaurant was filled, so we decided to go elsewhere.

这家餐馆已经没餐位了,我们得去别处了。

Specialized Sentences

1. Robots are becoming an increasingly prevalent adjunct in factories and industrial plants throughout the developed world.

在发达国家的工厂和车间里,机器人充当助手越来越普及。

2. Robots are programmed and engineered mechanical manipulators designed to perform industrial tasks without human intervention.

机器人是经设计并制造出来的机械手,可以独立从事各种工业活动。

3. Most of today's robots are employed in the automotive industry.

当今的机器人大多用于汽车工业。

4. Robots are programmed to take over such assembly line operations as welding and spray painting automobile and truck bodies.

机器人代替人类从事装配线上的操作,诸如给汽车和卡车焊接、喷漆等。

5. Robots already taking over human tasks in the automotive field.

在汽车行业,机器人已经代替人类做许多事情。

6. Robots are beginning to be seen, although to a lesser degree in other industries as well.

机器人在其他行业也开始出现,尽管还不是很常见。

7. Robots are more flexible and adaptable and usually more transportable than other machines.

机器人更灵活,适应性更强,并且通常比其他机器更易于移动。

8. We're going to transition into an economy that is very productive but that just doesn't need a lot of human workers.

我们将进入一个生产力非常高却不需要很多人类员工的经济时代。

9. AI has moved on from being seen as an overambitious and under-achieving field of research.

人工智能已从被视作预期过高而成就很低的研究领域摆脱出来。

10. The expression provided an attractive but informative name for a research program that encompassed such previously disparate fields as operations research, cybernetics, logic and computer science.

这种表述为研究计划提供了一个引人入胜且信息丰富的名称,该研究计划涵盖了诸如运筹学、控制论、逻辑学和计算机科学等不同的研究领域。

11. By the late 1980s, the term AI was being avoided by many researchers.

在 20 世纪 80 年代末,很多研究人员避免使用人工智能(AI)这一术语。

12. In particular, the problem of information overload, exacerbated by the growth of e-mail and the explosion in the number of web pages.

特别要指出的是信息负载、电邮数量激增以及网页数量急剧增加等问题。

13. AI can now be judged by what it can do, rather than by how well it matches up to a 30-year-old science-fiction film.

人们现在能根据人工智能所做的工作而不是通过它与一个 30 年前的科幻电影有多吻合来判断它的功能。

14. There are a variety of definitions of the term industrial robot.

关于工业机器人的定义有很多。

15. The number of robot installations worldwide varies widely.

世界各地机器人的数量会发生很大的变化。

16. Numerous single-purpose machines are used in manufacturing plants that might appear to be robots.

在制造工厂中使用的许多单一用途机器可能看起来像机器人。

17. These machines can only perform a single function and cannot be reprogrammed to perform a different function.

这些机器只有单一的功能,不能通过重新编程的方式完成不同的工作。

18. Such single-purpose machine is do not fit the definition for industrial robots that is becoming widely accepted.

这种单一用途的机器不能满足被人们日益广泛接受的关于工业机器人的定义。

19. An industrial robot is defined by the International Organization for Standardization (ISO) as an automatically controlled, reprogrammable, multipurpose manipulator.

国际标准化组织(ISO)对工业机器人的定义为:一种能够自动控制的、可重复编程的多功能机械手。

20. There exist several other definitions too, given by other societies, e. g. , by the Robot Institute of America (RIA), the Japan Industrial Robot Association (JIRA), British Robot Association (BRA) , and others.

其他一些协会,如美国机器人协会(RIA)、日本工业机器人协会(JIRA)、英国机器人协会(BRA)等,都对工业机器人提出了各自的定义。

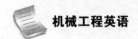

21. A robot is a reprogrammable multifunctional manipulator designed to move material, parts, tools, or specialized device.

机器人是一种用于移动材料、零件、工具或者专用装置的可重复编程的多功能机械手。

22. All definitions have two points in common. They all contain the words reprogrammable and multifunctional.

在所有的这些定义中有两个共同点,即可重复编程和多功能。

23. The term "reprogrammable" implies two things: the robot operates according to a written program.

"可重复编程"意味着两件事:机器人根据编写的程序工作。

24. This program can be rewritten to accommodate a variety of manufacturing tasks.

通过重新编写程序使其适应不同种类制造工作的需要。

25. The term "multifunctional" means that the robot can, through reprogramming and the use of different end-effectors, perform a number of different manufacturing tasks.

"多功能"意味着机器人能通过重复编程和使用不同的末端执行器来完成不同的制造工作。

26. The first articulated arm came about in 1951 and was used by the U.S. Atomic Energy Commission.

第一个关节式手臂于 1951 年被研制出来,供美国原子能委员会使用。

27. In 1954, the first industrial robot was designed by George C. Devol.

1954 年,第一个可以编程的机器人由乔治·C.沃德尔设计出来。

28. This is an unsophisticated programmable materials handling machine.

这是一台简单的可编程材料处理机器。

29. Are the droids taking our jobs?

机器人抢走了我们的工作吗?

30. And digital technologies are not just impacting knowledge work.

数字技术并非只对知识型的工作有影响。

31. Autonomous car, which is as cool as it sounds.

自动驾驶汽车,它就像它的名字一样酷。

32. The robots today still aren't very good at fixing bridges.

现在的机器人仍不擅长修建桥梁。

33. I'm still a huge digital optimist.

我是一个坚定的数字技术乐观主义者。

34. I am supremely confident that the digital technologies that we're developing now are going to take us into a utopian future.

我坚信我们现在正在开发的数字技术会带我们进入一个乌托邦社会。

35. The steam engine emerged during the industrial revolution.

蒸汽机出现在工业革命时期。

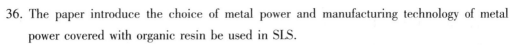

36. The paper introduce the choice of metal power and manufacturing technology of metal power covered with organic resin be used in SLS.

这篇文章主要介绍了用于 SLS 技术的覆膜金属粉的材料的选择和制备过程。

37. Digital technology these days, we are not anywhere near through with this journey.

从当今数码科技发展的趋势来看,我们还远未到达这趟旅程的终点。

38. Our digital—our technologies are great gifts.

我们的数码技术是伟大的礼物。

39. Right now, we have the great good fortune to be living at a time when digital technology is flourishing.

现在,我们有幸生活在数字技术蓬勃发展的时代

40. So, the droids are taking our jobs.

因此,机器人确实在取代我们的工作。

41. The point is that then we are freed up to do other things.

真正值得关注的是,技术带给我们自由从事其他活动的机会。

42. I am extremely confident for what we're going to do.

我对我们未来所做的很有信心。

43. Welcome our new computer overlords.

欢迎我们新的电脑霸主。

44. This is an automaton, a thinking machine.

这是一台自动机,一台会思考的机器。

45. These mechanical performers were popular throughout Europe.

这类机械表演曾经风靡欧洲。

46. That's robotic engineering in a pre-electronic age.

那是电子时代之前的机器人工程学。

47. The machines were far in advance of anything that Victorian technology could create.

这些机器远超前于维多利亚时代的技术。

48. Robots cannot anticipate human actions.

机器人不能预测人类的动作。

49. You can imagine that electrons are kind of bees moving around, buzzing around this nucleus.

你可以想象电子就像蜜蜂一样在原子核周围移动、转圈。

50. Why human and robots find it difficult to work in close proximity?

为什么人类和机器人很难近距离地工作?

Task Two   Sample Dialogue

Directions: *In this dialogue, you are going to read several times the following sample dialogue about the relevant topic. Please pay special attention to five Cs (culture, context, coherence, cohesion and critique) in the dialogue and get ready for a smooth communication in the coming task.*

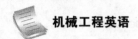

Visiting an Industrial Robot Show

(*Tom and Mary are now visiting an industrial robot show in a university. A young girl named Marcy is in charge of showing the visitors around.*)

Marcy: Ladies and gentlemen, welcome to our industrial robot show. I'm Marcy Margaret. My job is to show you around. Please don't hesitate to ask questions.

Mary: How interesting to see so many kinds of robots here. Excuse me, Marcy, but I wonder why these robots do not look like nor behave like human beings?

Marcy: Ah, I see. The point here is that what we show you are not other kinds of robots but industrial ones. We needn't make them look like us humans.

Mary: Well, I understand. Now could you tell us what these industrial robots can do for us?

Marcy: Sure. These robots can do many things for us. They can help us, for example, to handle materials, spray paint, rescue people from the fire, and even help to explore deep oceans and outer-space.

Tom: Uh, how great robots are! But what makes robots able to work for us so marvelously?

Marcy: It's nothing else but the computer programs. Robots are a reprogrammable mechanical manipulator. Being programmed with human instructions, they can move along several directions and do factory work usually done by human beings.

Mary: Oh, that's wonderful. But what's the difference between robot and my handy calculator or my home washing machine, now that all of them belong to some reprogrammable mechanical manipulator.

Marcy: Okay, it's a good question. But I think, here you mix a robot with an ordinary washing machine because you forget a robot can do a stand-alone operation, that is, rather independent jobs, while a washing machine can't. So, when we talk about industrial robots, we mean a machine or device, which is something reprogrammable with its own end effector, which can perform a factory duty in a stand-alone manner.

Tom: Oh, I see. I guess, my automatic watch and my home washing machine cannot play chess with me while a robot can.

Marcy: That's right. You're a smart boy, I see.

Task Three   Simulation and Reproduction

Directions: *The class will be divided into three major groups, each of which will be assigned a topic. In each group, some students may be the teacher, while others may be students. In the process of discussion, please observe the principles of cooperation, politeness and choice of words. One of the groups will be chosen to demonstrate the discussion to the class.*

1) application of industrial robots in our life
2) a funny story related to robots in my life

3) What new development in industrial robots do you expect to see in the near future?

Task Four   Discussion and Debate

Directions: *The class will be divided into two groups. Please choose your stand in regard to the following controversy and support your opinions with scientific evidences. Please refer to the specialized terms and classical sentences in the previous parts of this unit.*

These robots do not look or behave like human beings, but they do the work of humans. Robots are particularly useful in a wide variety of applications, such as material handling, spray painting, spot welding, arch welding, inspection, and assembly. Which party do you agree with? Why? Are the robots taking our jobs?

V. After-class Exercises

1. *Match the English words in Column A with the Chinese meaning in Column B.*

A	B
1) hydraulic	A) 焊接
2) actuator	B) 装配
3) axis	C) 制动器
4) assembly	D) 轴
5) welding	E) 液压的
6) spray	F) 腕关节
7) gripper	G) 可编程
8) jaw	H) 手爪
9) wrist	I) 爪钳夹装置
10) programmable	J) 喷

2. *Fill in the following blanks with the words or phrases in the word bank. Change the forms if it's necessary.*

robot	flexible	manipulator	1951	end-effectors
George C. Devol	industrial	automotive	mulitipurpose	human

1) An industrial robot is defined by the International Organization for Standardization (ISO) as an automatically controlled, re-programmable, _____ manipulator.

2) Robots already taking over _____ tasks in the automotive field.

3) Robots are more _____ and adaptable and usually more transportable than other machines.

4) A robot is a reprogrammable multifunctional _____ designed to move material, parts, tools, or specialized devices through variable programmed motions for the performance of a variety of tasks.

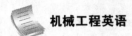

5) The first articulated arm came about in _____ and was used by the U.S. Atomic Energy Commission.

6) In 1954, the first industrial robot was designed by _____.

7) Robots, becoming an increasingly prevalent adjunct in factories and _____ plants.

8) Most of today's robots are employed in the _____ industry, where they are programmed to take over such assembly line operations as welding and spray painting automobile and truck bodies.

9) It is not yet known whether _____ will one day have vision as good as human vision.

10) The term "multifunctional" means that the robot can, through reprogramming and the use of different _____, perform a number of different manufacturing tasks.

3. *The following is the main structure of a industrial robots (a mechanical gripper for clamping cylindrical materials). Can you write the corresponding parts?*

1 _____
2 _____
3 _____
4 _____
5 _____

4. *Translate the following sentences into English.*

1) "多功能"整个词意味着机器人通过重复编程和使用不同的末端执行器,来完成不同的制造工作。

   _____

2) 我们将进入一个生产力非常高却不需要很多人类员工的时代。

   _____

3) 人工智能已从被视作预期过高,而成就很低的研究领域走了出来。

   _____

4) 机器人更灵活,更有适应性,并且通常比其他机器更易于移动。

   _____

5）从当今的经济和社会发展趋势来看，我认为我们尚未见证真正的变革。未来可期。

_____

_____

5. *Please write an essay of about* 120 *words on the topic*：***Application of industrial robots in our life.*** *Some specific examples will be highly appreciated and you have to watch out the spelling of some specialized terms you have learnt in this unit.*

_____

_____

_____

_____

_____

_____

_____

_____

_____

_____

_____

VI. Additional Reading

The History and Organizational Structure of ABB Group

ABB（ASEA Brown Boveri）is a Swedish-Swiss multinational corporation（跨国公司）headquartered（总部）in Zurich（苏黎世），Switzerland（瑞士），operating mainly in robotics，power，heavy electrical equipment（重型电气设备）and automation technology areas. It is ranked 341st in the Fortune 500 global list of 2018 and has been a global Fortune 500 company for 24 years.

ABB is traded on the SIX Swiss Exchange in Zurich，Nasdaq Stockholm and the New York Stock Exchange（证券交易所）in the United States.

ABB's history goes back to the late 19th century. Allmänna Svenska Elektriska Aktiebolaget（General Swedish Electrical Limited Company，ASEA）（瑞典通用电气有限公司）was founded in 1883 by Ludvig Fredholm in Västerås as a manufacturer of electrical light and generators（发电机）. Brown，Boveri & Cie（BBC）was formed in 1891 in Baden，Switzerland，by Charles Eugene Lancelot Brown and Walter Boveri as a Swiss group of electrical companies producing AC and DC motors（交流和直流电机），generators，steam turbines（蒸汽轮机）and transformers（变压器）.

ABB was created as the result of the merger（合并）of the Swedish corporation ASEA and the

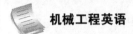
Swiss company Brown, Boveri & Cie (BBC) in 1988. The latter had acquired (收购) Maschinenfabrik Oerlikon in 1967. The former CEO of ASEA, Percy Barnevik ran the company until 1996.

ABB Around the World

ABB bought International Combustion Ltd from Rolls-Royce in 1997. In 2011, ABB acquired Baldor Electric USA for $4.2 billion in an all-cash transaction On January 30, 2012, ABB Group acquired Thomas & Betts in a $3.9 billion cash transaction. On June 15, 2012, it completed the acquisition of commercial and industrial wireless technology specialists Tropos. In July 2013, ABB acquired Power-One in a $1 billion all-cash transaction, to become the leading global manufacturer of solar inverters. On June 30, 2018, ABBcompleted its acquisition (收购) of GE Industrial Solutions, GE's global electrification business(GE全球电气化业务). The transaction (交易) was announced (公布) on September 25, 2017.

ABB Transformer in Iowa

The Robotics and Motion(运动) division provides products and services for industrial production. It includes electric motors (电动机), generators, drives(驱动器), power electronics(电力电子) and industrial robots. ABB has installed (安装) over 300,000 robots. In 2006, ABB opened a manufacturing center in Shanghai, China. Also, wind(风力) generators, solar power inverters(逆变器) and UPS products belong to this division.

An ABB Industrial Robot

The Industrial Automation division provides systems for control, plant optimization(工厂优化), and industry-specific automation applications(行业特定自动化应用系统). The industries served include oil and gas (天然气), power (电力), chemicals and pharmaceuticals(制药), pulp (纸浆) and paper, metals(金属) and minerals, marine(船舶) and turbocharging(涡轮增压). The division consists of seven business units: Control Technologies (the world's No.1 DCS supplier); Marine & Ports; Measurement & Analytics; Oil, Gas & Chemicals; Power

Generation & Water；Process Industries(流程工业) and Turbocharging.

(*If you want to find more information about this corporation，please log on https：//en. wikipedia. org/wiki/ABB_Group.*)

1. *Read the passage quickly by using the skills of skimming and scanning，and choose the best answer to the following questions.*

1) Where is the ABB's headquarter in Switzerland? _____

    A. Geneva   B. Basel

    C. Lyss   D. Zurich

2) It is ranked _____ in the Fortune 500 global list of 2018.

    A. 28th   B. 500th

    C. 341st   D. 300th

3) ABB is traded on the six Swiss Exchange in Zurich，Nasdaq Stockholm and the _____ Stock Exchange(证券交易所) in the United States.

    A. Los Angeles   B. Washington D. C.

    C. New York   D. Houston

4) ABB's history goes back to the _____ century.

    A. late 19th   B. early 19th

    C. late 20th   D. early 20th

5) ABB was created as the result of the merger(合并) of the Swedish corporation ASEA and the Swiss company Brown，Boveri & Cie (BBC) in _____.

    A. 1996   B. 1988

    C. 1997   D. 2017

6) ABB bought International Combustion Ltd.from _____ in 1997.

    A. BMW   B. Rolls-Royce

    C. Audi   D. Baldor Electric USA

7) On January 30，2012，ABB Group acquired Thomas & Betts in a _____ cash transaction.

    A. $4.2 billion   B. $4.2 million

    C. $3.9 million   D. $3.9 billion

8) When does ABB completed its acquisition(收购)of GE Industrial Solutions? _____

    A. June 30，2018   B. September 25，2017

    C. January 30，2012   D. June 15，2012

9) In 2006，ABB opened a manufacturing centre in _____，China.

    A. Beijing   B. Guangzhou

    C. Shanghai   D. Shenzhen

10) The division consists of _____ business units.

    A. seven   B. eight

    C. nine   D. ten

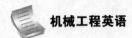

2. *In this part, the students are required to make an oral presentation on either of the following topics.*

1) the history of ABB

2) the lessons from ABB's development history

习题答案

Unit Nine Modern Design

I. Pre-class Activity

Directions: *Please read the general introduction about **John Walker** and tell something more about the great computer programmer to your classmates.*

John Walker is a computer programmer (程序员), author and co-founder of the computer-aided design software company(计算机计算辅助设计软件公司) Autodesk. Walker also founded the hardware integration manufacturing company(硬件集成制造公司) Marinchip. Among other things, Marinchip pioneered the translation of numerous computer language compilers(编译器) to Intel platforms.

In 1982, John Walker and 12 other programmers pooled (汇集) US $59,000 to start Autodesk (Auto CAD), and began working on several computer applications(计算机应用程序). The first completed was Auto CAD, a software application for computer-aided design (CAD) (计算机辅导设计) and drafting. Auto CAD had begun life as Interact CAD, written by programmer Michael Riddle in a proprietary language. Walker and Riddle rewrote the program, and established a profit-sharing agreement for any product derived from Interact CAD. Walker subsequently paid Riddle US $10 million for all the rights.

By mid−1986, the company had grown to 255 employees with annual sales of over $40 million. That year, Walker resigned (辞职) as chairman and president of the company, continuing to work as a programmer. In 1989, Walker's book, *The Autodesk File*, was published. It describes his experiences at Autodesk, based around internal documents (particularly email) of the company.

Walker moved to Switzerland in 1991. By 1994, when he resigned from the company, it was the sixth-largest personal computer software company in the world, primarily from the sales of Auto CAD. Walker owned about $45 million of stock in Autodesk at the time.

II. Specialized Terms

Directions: *Please remember the following specialized terms before the class so that you will be*

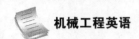

able to better cope with the coming tasks.

archaic adj. 古代的;陈旧的

Autodesk n. 美国电脑软件公司

avid adj. 渴望的,热心的

zeppelin n. 齐柏林硬式飞艇

bar n. 条

CD-ROM drive 光驱

cable-stayed bridge 斜拉桥

clibration curve 校准曲线

clibrator of accelerometer 加速计校准仪

cnonical coordinates 正则坐标

canonical distribution 正则分布

canonical equation of motion 正则运动方程

canonical form 正则形式

canonical momentum 正则动量

canonical transformation 正则变换

canonical variable 正则变量

centrifugal adj. 离心的

centrifugal force 离心力

centrifugal governor 离心式调速器

charge n. 费用;电荷;掌管;控告

check v. 检查;验收

chemical engineering 化学工程

chip vt. 削,凿;削成碎片 vi. 剥落

circulate v. 传播;流传;循环

citrus 柑橘属果树

civil engineering 土木工程

clip n. (塑料或金属的)夹子

cloud application 云应用

cloud computing 云计算

cloud server 云服务器,云端伺服器

cloud storage 云存储

coin vt. 创造

collaborate vi. 合作

drawing n. 图画

collectively adv. 共同的,集体的

combination n. 结合;组合

combined camera 结合相机

command vi. 命令,指挥;控制

communicate v. 通讯,传达

communication n. 通信

compasses n. 圆规;罗盘(compass 的复数)

component adj.组成的,构成的

dimensional adj. 空间的

drafting n. 起草;制图

drive n. 驱动器

druid n. 德鲁伊教团员

electrify v. 使电气化

electrode n.电极;电焊条

electromagnetic adj. 电磁的

electro-mechanical 机电的

electro-mechanics n. 机电学;电机机械

electron n. 电子

electronic adj. 电子的

electronic communication devices
电子通信装置

electronic computer 电子计算机

electronic signal 电子信号

element n. 因素

encompass n. 围绕;包含

essential adj. 基本的;必要的

excavate v. 发掘;挖掘

exterior equipment 外部设备,外接设备

extraterrestrial adj. 地球外的

feedback n. 反馈,回馈

feedback control systems 反馈控制系统

figure n. 数字;人物;图像;价格

file n. 文件;档案;文件夹;锉刀

file access 文件存取

fixed network 固定网络

fluid mechanics[流]流体力学,液体力学

format n. 格式;版式

formula n.[数]公式,准则

forward-thinking 前瞻性的

frame n.框架;结构;[电影]画面 vt. 设计;
建造;陷害;使……适合

function n. 功能;函数;职责;盛大的集会。

fundamental adj. 基本的

fusion n. 融合;融化;熔接

futuristic adj. 未来派的

generalized adj. 广义的

generate vt. 使形成;发生

icon n. 图标

iconic adj.图标的,形象的

interface n. 界面

imminent adj. 即将来临的;迫近的

methodology n.方法学

misnomer n. 用词不当;误称

modify vt. 修改

murky adj. 黑暗的;朦胧的;阴郁的

offset n. 抵消,补偿;平版印刷

panel n. 仪表板

philosophy n. 哲学;哲理

remark vt. 评论

ribbon n. 带;缎带

savagery n. 野性;野蛮人

seasoned adj. 经验丰富的

serpent n. 狡猾的人

shaped adj. 合适的

skull n. 头盖骨,脑壳

slightly adv. 略微地

subtle adj. 微妙的;精细的

synonymous adj. 同义的

tab n. 标签;制表

version n. 版本;译文

III. Watching and Listening

Task One   John Hodgman：Design, and its Interpretation

New Words

iconic adj. 图标的,形象的

avid adj. 渴望的,热心的

zeppelin n. 齐柏林硬式飞艇

murky adj. 地球外的

excavate v. 发掘;挖掘

dimensional adj. 空间的

savagery n. 野性;野蛮人;原始状态

druid n. 德鲁伊教团员

archaic adj. 古代的;陈旧的

futuristic adj. 未来派的

synonymous adj. 同义的

misnomer n. 用词不当;误称

serpent n. 狡猾的人

skull n. 头盖骨,脑壳

imminent adj. 即将来临的;迫近的

citrus n.[园艺]柑橘属果树;柑橘类的植
物 adj. 柑橘属植物的

视频链接及文本

Exercises

1. *Watch the video for the first time and choose the best answer to the following questions.*

    1) How many examples of iconic design in the text? _____

        A. three   B. half

        C. three and a half   D. two

    2) What education does the author have? _____

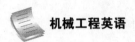

A. Bachelor's degree   B. Master's degree

C. Doctor's degree   D. Double degree

3）Which city does the author mentioned？_____

  A. New York   B. Los Angeles

  C. Las Vegas   D. Washington

4）What was the first design mentioned in the article？_____

  A. iPhone   B. Juicy Salif

  C. alien   D. Theme Building

5）Which iPhone features the author won't use？_____

  A. sell books   B. read books

  C. buy books   D. measure the weight

2. *Watch the video again and decide whether the following statements are true or false.*

1）It was first excavated in 1961 as they were building LAX. (　)

2）Today I'm going to unpack for you two examples of iconic design. (　)

3）Juicy Salif is a design by Philippe Starck. (　)

4）This is affordable and can come home with you and, as such, it can sit on your kitchen counter. (　)

5）Unlike the Theme Building, this is alien technology.(　)

3. *Watch the video for the third time and fill in the following blanks.*

One _____ thing about it , if you do have one at _____, let me tell you one of the _____ you may not know: when you fall _____, it comes alive and it walks around your house and goes _____ your mail and watches you as you sleep. Okay, what is this _____? I have no idea. I don't know what that thing is. It looks _____. Is it a little hot _____? I don get it. Does anyone know? Chi? It's an... iPhone. iPhone. Oh yes, that's right, I remember those; I had my whole bathroom tiles _____ with those back in the good old days. No, I have _____ iPhone. Of course I do.

4. *Share your opinions with your partners on the following topics for discussion.*

1）Do you like iPhone? What modern design do you like most ? Please summarize the features of products.

2）Can you use a few lines to list what's your understanding about design? Please use an example to clarify your thoughts.

Task Two   Auto CAD 2019 Essential Training

New Words

essential adj. 基本的；必要的   methodology n. 方法学

philosophy n.哲学；哲理   drive n. 驱动器

seasoned adj.经验丰富的，老练的   drafting n. 起草；制图

视频链接及文本

version n. 版本;译文

format n. 格式;版式

subtle adj. 微妙的;精细的

interface n. 界面

Autodesk n. 美国电脑软件公司

icon n. 图标

panel n. 仪表板

tab n. 标签;制表

modify vt. 修改

slightly adv. 略微地

shaped adj. 合适的

remark vt. 评论

offset n. 抵消,补偿;平版印刷

ribbon n. 带;缎带

collaborate vi. 合作

drawing n. 图画

commands n.[计]命令

features n. 产品特点

bar n. 条

couple n. 对;夫妇

Exercises

1. *Watch the video for the first time and choose the best answer to the following questions.*

   1) The whole idea of this course is it'll teach you all the _____ and naming philosophies.

      A. methodologies   B. training

      C. Auto CAD   D. methods

   2) The first thing you should know is that we are using the _____ version of Auto CAD.

      A. newest   B. latest

      C. quickest   D. fastest

   3) The _____ you are when you're drafting with Auto CAD, the quicker you are in the office.

      A. more   B. less

      C. quicker   D. slower

   4) Now, as mentioned in what you should know about this _____ course, we're using the DWG file format.

      A. old   B. new

      C. special   D. particular

   5) Now, _____ is a new feature in Auto CAD 2019.

      A. home tab   B. new icon

      C. drawing compare   D. quick access

2. *Watch the video again and decide whether the following statements are true or false.*

   1) What we're going to be doing as we work through this course is getting you up to speed with Auto CAD and making you a seasoned Auto CAD user. (   )

   2) The whole idea is that you want to waste time when you're using Auto CAD. (   )

   3) That's the whole idea of these courses, to make you better and empower you and to drive you forward using Auto CAD. (   )

   4) You can open from the Auto CAD web-based version and the Auto CAD mobile-based version. (   )

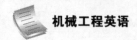

5) They're very hard changes, very subtle changes, but you might not notice them otherwise. ( )

3. *Watch the video for the third time and fill in the following blanks of the table.*

Features	Auto CAD 2013	Auto CAD 2019
1		
2		
3		

4. *Share your opinions with your partners on the following topics for discussion.*
   1) What are the methodologies of CAD?
   2) operation and application of CAD in modern design

IV. Talking

Task One　Classical Sentences

Directions：*In this section, some popular sentences are supplied for you to read and to memorize. Then, you are required to simulate and produce your own sentences with reference to the structure.*

General Sentences

1. My hobby is collecting stamps. Do you have a hobby?
   我的爱好是集邮。你有什么爱好？

2. I've always thought photography would be an interesting hobby.
   我一直认为摄影是一种有趣的爱好。

3. Some people like horseback riding, but I prefer golfing as a hobby.
   有些人喜欢骑马,但是我喜欢打高尔夫。

4. Do you have any special interests other than your job?
   除了工作以外,你还有什么其他特殊爱好吗？

5. Learning foreign languages is just an avocation with me.
   学外语只是我的业余爱好。

6. I find stamp collecting relaxing and it takes my mind off my work.
   我发现集邮使人放松,能让我的注意力从工作中移开。

7. On weekends I like to get my mind off my work by reading good books.
   周末我喜欢通过读好书把注意力从工作上转移开。

8. My cousin is a member of a drama club. He seems to enjoy acting.
   我堂兄是戏剧俱乐部的一名成员,他似乎喜欢表演。

9. He plays the piano for his own enjoyment.
他弹钢琴是为了自娱自乐。

10. I've gotten interested in WiFi. I'm building my own equipment.
我对无线网络感兴趣,我正在安装我自己的设备。

11. He's not a professional. He plays the piano for the fun of it.
他不是专业人士,他弹钢琴是为了好玩。

12. I've heard of unusual hobbies, but I've never heard of that one.
我听说过一些不寻常的爱好,但我从来没听说过那一个。

13. The trouble with photography is that it's an expensive hobby.
摄影的问题在于,它是一种昂贵的爱好。

14. That's a rare set of coins. How long did it take you to collect them?
这是一套罕见的钱币。你用了多长时间收集的?

15. I started a new hobby. I got tired of working in the garden.
我开始了一个新的爱好,我厌倦了在花园里干活。

16. Baseball is my favorite sport. What's your favorite?
棒球是我最喜欢的体育运动。你最喜欢的是什么?

17. My nephew is a baseball player. He is a catcher.
我的外甥是一名棒球运动员,他是一名接球手。

18. When you played football, what position did you play?
你踢足球时, 踢什么位置?

19. We played a game last night. The score was tired six-to-six.
昨晚我们玩了一场比赛,比分是 6 比 6 平。

20. I went to a boxing match last night. It was a good fight.
昨晚我看了一场拳击比赛,比赛很精彩。

21. When I was on the track team, I used to run the quarter mile.
我在田径队时,经常跑四分之一英里。

22. I like fishing and hunting, but I don't like swimming.
我喜欢钓鱼和打猎,但是不喜欢游泳。

23. My favorite winter sport is skiing. I belong to a ski club.
我最喜欢的冬季运动是滑雪。我参加了一个滑雪俱乐部。

24. Would you be interested in going to the horse races this afternoon?
今天下午你有兴趣去看赛马吗?

25. What's your favorite kind of music? Do you like jazz?
你最喜欢哪种音乐类型? 你喜欢爵士乐吗?

26. He's a composer of serious music. I like his music a lot.
他是一名严肃音乐作曲家,我很喜欢他的音乐。

27. My brother took lessons on the trumpet for nearly ten years.
我的哥哥练习吹喇叭将近十年了。

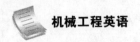

28. You play the piano beautifully. How many hours do you practice every day?
    你钢琴弹得很好,你每天练习多长时间?

29. I've never heard that piece before. Who wrote it?
    我从没有听过这一段,是谁写的?

30. Have you ever thought about becoming a professional musician?
    你有没有想到过要成为一名专业的音乐家?

31. Be a good sport. Play according to the rules of the game.
    成为一个好的运动员需要遵守游戏规则。

32. Our family went camping last summer. We had to buy a new tent.
    我家去年夏天去露营了。我们得买个新的帐篷。

33. This afternoon we went to the gym for a workout. We lifted weights.
    今天下午我们去体育馆健身了,我们练了举重。

34. What do you do for recreation? Do you have a hobby?
    你闲暇时做什么? 有什么爱好吗?

35. My muscles are sore from playing baseball.
    打完棒球后,我的肌肉一直酸痛。

36. I sent in a subscription to that magazine. It's put out every week.
    我订阅了那份杂志,它是周刊。

37. If you subscribe to the newspaper, it'll be delivered to your door.
    如果你订阅报纸的话,可以送到你家。

38. I didn't read the whole paper. I just glanced at the headlines.
    我没有通读全文,只是看了看标题。

39. The first chapter of the story is in this issue of the magazine.
    这个故事的第一章刊登在本期杂志上。

40. I haven't seen the latest issue of the magazine. Is it out yet?
    我还没有看到这个杂志的最新一期,是不是还没有出版?

41. What's the total circulation of this newspaper?
    这个报纸的总发行量怎样?

42. I'm looking for the classified section. Have you seen it?
    我在找分类广告栏,你看到了吗?

43. My brother-in-law is a reporter on *The New York Times* staff.
    我姐夫是《纽约时报》的记者。

44. There was an article in today's paper about the election.
    今天的报纸上有选举的消息。

45. There wasn't much news in the paper today.
    今天的报纸上没有太多的消息。

46. How long have you been taking this magazine?
    你订这份杂志多久了?

47. Did you read the article about the rescue of the two fishermen?

你读了那篇关于营救两名渔夫的文章了吗?

48. Why don't you put an advertisement in the paper to sell your car?

你为什么不在报纸上登个卖车广告呢?

49. I got four replies to my ad.about the bicycle for sale.

我的自行车待售广告有四个回复。

50. My son has a newspaper route. He delivers the morning paper.

我儿子有送报的路线,他送晨报。

Specialized Sentences

1. Computer-aided design(CAD) is the use of computer technology to aid in the design and particularly the drafting of a part or product.

计算机辅助设计(CAD)是利用计算机技术辅助设计,特别是辅助零件或产品的绘图技术。

2. Nowadays, CAD has become an especially important technology, which greatly improves work efficiency compared with hand drafting.

如今,计算机辅助设计已经成为一项特别重要的技术,与手工绘图相比,它大大提高了工作效率。

3. Today I'm going to unpack for you three examples of iconic design.

今天我将为大家展示三个标志设计的例子。

4. CAD enables designers to lay out and develop work on screen, print it out and save it for future editing, thus saving time on their drawings.

CAD 可以让设计师在屏幕上布置和开发工作,打印出来并保存以备日后编辑,从而节省了绘制图纸的时间。

5. There are several different types of CAD.

CAD 有几种不同的类型。

6. Each type requires the operator to think differently about how he will use it and design virtual components in a different manner.

每种类型都要求操作员以不同的方式考虑如何使用它,并以不同的方式设计虚拟组件。

7. There are many producers of the lower-end 2D systems.

有许多低端2D 系统的生产商。

8. These provide an approach to the drawing process without the trouble on scale and placement on the drawing sheet that accompanied hand drafting, since these can be adjusted as required during the creation of the final draft.

这为绘图过程提供了一种方法,不需要在绘图页上的标度和位置上加上手工绘图,因为在创建最终草稿时,这些可以根据需要进行调整。

9. With their help, designers do not need to worry about scale and placement on the drawing

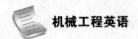

sheet.

在他们的帮助下,设计师们不需要担心图纸上的比例和位置。

10. 3D wireframe is basically an extension of 2D drafting.

三维线框图基本上是二维制图的延伸。

11. Each line has to be manually inserted into the drawing.

每一行都必须手工插入图中。

12. The final product has no mass properties associated with it and cannot have features directly added to it.

最终产品没有与之相关的质量属性,也不能直接添加特性。

13. 3D "dumb" solids are created in a way similar to controlling real world objects.

3D"哑"实体的创建方式类似于控制现实世界的物品。

14. Basic three-dimensional geometric forms have solid volumes added or subtracted from them, as if assembling or cutting real world objects.

基本的三维几何形状有实体的增加或减少,就像组装或切割真实世界的物体。

15. Two-dimensional projected views can easily be generated from the models.

二维投影视图可以很容易地从模型中生成。

16. 3D parametric solid modeling requires the operator to use what is referred to as "design intent".

三维参数实体建模要求操作者使用所谓的"设计意图"。

17. The objects and features created are adjustable.

创建的物体和特性是可调的。

18. Any future modifications will be simple, difficult, or nearly impossible, depending on how the original part was created.

未来的任何修改都将是或难或易,甚至是几乎不可能的,这取决于最初的部分是如何创建的。

19. If a feature was intended to be located from the center of the part, the operator needs to locate it from the center of the model, not, perhaps, from a more convenient edge or any point, as he could when using "dumb" solids.

如果一个特性要从部件的中心定位,操作员需要从模型的中心定位它,而不是像使用"哑"实体时那样,从更方便的边缘或任何点定位它。

20. Parametric solids require the operator to consider the consequences of his actions carefully.

参数实体要求操作者仔细考虑其行为的后果。

21. In principle, CAD could be applied throughout the design process, but in practice its impact on the early stages has been limited because sketches are widely used at that time.

原则上,CAD 可以应用于整个设计过程,但实践中,由于草图在当时被广泛使用,所以它对早期阶段的影响有限。

22. There are some new software programs currently available for these stages.

目前有一些新的软件可用于这些阶段。

23. It remains to be seen how effective they will be and how widely they will be implemented.

这些措施的效果如何,实施范围有多大,还有待观察。

24. Computer-aided manufacturing (CAM) can be defined as the use of computer systems to plan, manage, and control the operations of a manufacturing plant through either direct or indirect computer interface with the plant's production resources.

计算机辅助制造(CAM)可以定义为使用计算机系统,通过与生产厂家的生产资源直接或间接的计算机接口来计划、管理和控制生产厂家的操作。

25. CAM functions center around four main areas: numerical control, process planning, robotics, and factory management.

CAM 功能主要集中在四个领域:数控、工艺规划、机器人技术和工厂管理。

26. In modern CNC systems, end-to-end component design is highly automated using CAM programs.

在现代 CNC 系统中,通过使用 CAM 程序端到端组件设计是高度自动化的。

27. The programs produce a computer file that is interpreted to extract the commands needed to operate a particular machine, and then loaded into the CNC machines for production.

程序生成一个计算机文件,该文件被解释为提取操作特定机器所需的命令,然后加载到 CNC 机器进行生产。

28. Process planning is involved with the detailed sequence of production steps from start to finish.

过程计划包括从开始到结束的生产步骤的详细顺序。

29. Essentially, the process plan describes the state of the workplace at each workstation.

从本质上讲,过程计划描述了每个工作站工作场所的状态。

30. The use of computers as an aid to process planning is comparatively recent and has led to a rebirth of what is known as group technology (GT).

使用计算机辅助过程规划的历史相对较晚,导致所谓成组技术(GT)的重生。

31. Group technology is based on organizing all similar parts into families to allow standardization of manufacturing steps.

成组技术的基础是将所有相似的零件归为一组,以实现制造步骤的标准化。

32. Currently a process planning system is under development and it is able to produce process plans directly from the geometric model database with almost no human assistance.

目前正在开发的过程计划系统可以直接从几何模型数据库中生成过程计划,几乎不需要人工协助。

33. In this system, the process planner would review the impact from the design engineer via communication and then enter this input into the CAM system which would generate a complete set of process plans automatically.

在这个系统中,过程计划者通过沟通来审查设计工程师的影响,然后将此影响输入 CAM 系统,CAM 系统会自动生成一套完整的过程计划。

34. Many advances are being made to integrate robotics into CAM.

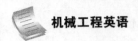

在将机器人技术集成到 CAM 方面取得了许多进展。

35. One of these efforts is the US Air Force integrated computer-aided manufacturing (ICAM) project, of which the goal is to organize every step of manufacturing.

其中之一是美国空军综合计算机辅助制造(ICAM)项目,其目标是组织生产的每一个步骤。

36. In engineering, design is a component of the engineering process.

在工程学中,设计是工程过程的组成部分之一。

37. Many overlapping methods and processes can be seen when comparing product design, industrial design and engineering design.

在比较产品设计、工业设计和工程设计时,可以看到许多重叠的方法和过程。

38. "To formulate a plan", and defines engineering as: "The application of scientific and mathematical principles to practical ends such as the design, manufacture, and operation of efficient and economical structures, machines, processes, and systems."

"制定计划",并将工程定义为:"将科学和数学原理应用于实际目的,如高效、经济的结构、机器、过程和系统的设计、制造和运行。"

39. He led the study of the test, called Corus CAD.

他领导了这项称为康丽斯 CAD 的测试研究。

40. We want to make it mainstream by targeting CAD-like productivity improvements.

通过以类 CAD 的生产力改进为目标,我们想使它成为主流。

41. You can now open it from the web and from mobile, so you can open it from the Auto CAD web-based version and the Auto CAD mobile-based version.

现在您可以从网页和移动平台上打开它,因此您也可以从基于网页的 Auto CAD 版本和基于移动平台的 Auto CAD 版本中打开它。

42. How much science is applied in a design is a question of what is considered "science".

在设计中应用了多少科学被认为是"科学"问题。

43. Along with the question of what is considered science, there is social science versus natural science.

除了什么被认为是科学这一问题外,还有什么是社会科学和自然科学。

44. Scientists at Xerox PARC made the distinction of design versus engineering at "moving minds" versus "moving atoms".

施乐帕洛阿尔托研究中心(Xerox PARC)的科学家们对"移动的头脑"和"移动的原子"的设计和工程进行了区分。

45. So there's just a couple of new little features in there that are quite cool and quite useful in Auto CAD 2019.

在 Auto CAD 2019 中有一些很酷很有用的新功能。

46. The relationship between design and production is one of planning and executing.

设计与生产的关系是计划与执行的关系。

47. In theory, the plan should anticipate and compensate for potential problems in the execution process.

从理论上讲,该计划应该预见并补偿执行过程中潜在的问题。

48. In contrast, production involves a routine or pre-planned process.

相反,生产涉及一个常规的或预先计划好的过程。

49. A design may also be a mere plan that does not include a production or engineering processes although a working knowledge of such processes is usually expected of designers.

设计可能仅仅是一个不包含生产或工程过程的计划,尽管设计师通常需要对这些过程的工作知识有所了解。

50. In some cases, it may be unnecessary or impractical to expect a designer with a broad multidisciplinary knowledge required for such designs to also have a detailed specialized knowledge of how to produce the product.

在某些情况下,期望一个拥有广博多学科知识的设计师对如何生产产品有详细的专业知识是不必要的或不切实际的。

Task Two   Sample Dialogue

Directions: *In this dialogue, you are going to read several times the following sample dialogue about the relevant topic. Please pay special attention to five Cs (culture, context, coherence, cohesion and critique) in the dialogue and get ready for a smooth communication in the coming task.*

CAD and CAM are Widely Applied

(*Jessica and Andrew are talking about CAD and CAM are widely applied in mold design and mold making.*)

Jessica: Mr. Johnson, I learned that CAD and CAM are widely applied in mold design and mold making. Can you tell me something more about them?

Andrew: Sure. CAD is the use of computer technology for the design of objects. And CAM is a system of using computer technology to assist the manufacturing process.

Jessica: Then how are these two systems related?

Andrew: CAM is commonly linked to CAD systems. The integrated CAD/CAM system takes the computer design and puts it directly into the manufacturing system; the design is then turned into many computer-controlled processes, such as drilling.

Jessica: CAM is one of the most dramatic changes in the manufacturing process since the Industrial Revolution, isn't it?

Andrew: Exactly.

Jessica: And I also knew there are four main areas of CAM. But I don't remember what they are exactly.

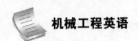

Andrew: You mean numerical control, process planning, robotics, and factory management?

Jessica: Yes, that's right! Could you please tell me more about them?

Andrew: Definitely! All these systems are concerned with a highly automated factory. Because each of the manufacturing processes in a CAM system is computer controlled, a high degree of precision can be achieved, which is impossible with manual manufacturing.

Jessica: So CAM can raise the production rates?

Andrew: Yes. It allows a company to get the best from workers by increasing their productivity. It also has many other advantages.

Jessica: For example?

Andrew: CAM allows the process planner to receive the impact from the design engineer via communication. It helps coordinate operations of an entire factory.

Jessica: That sounds great! Thank you so much, Mr. Johnson, I learned a lot today.

Task Three　Simulation and Reproduction

Directions: *The class will be divided into three major groups, each of which will be assigned a topic. In each group, some students may be the teacher, while others may be students. In the process of discussion, please observe the principles of cooperation, politeness and choice of words. One of the groups will be chosen to demonstrate the discussion to the class.*

1) modern design in our daily life
2) What are the functions of CAD?
3) the importance of learning CAD

Task Four　Discussion and Debate

Directions: *The class will be divided into two groups. Please choose your stand in regard to the following controversy and support your opinions with scientific evidences. Please refer to the specialized terms and classical sentences in the previous parts of this unit.*

CAD and CAM are widely applied in mold design and mold making. Which party do you like most? Why?

V. After-class Exercises

1. *Match the English words in Column A with the Chinese meaning in Column B.*

A	B
1) dimensional	A) 图标的
2) futuristic	B) 空间的
3) synonymous	C) 同义的

4）iconic   D）格式

5）philosophy   E）仪表板

6）methodology   F）方法学

7）drive   G）哲学

8）version   H）未来主义的

9）format   I）版本

10）panel   J）驱动器

2. *Fill in the following blanks with the words or phrases in the word bank. Change the forms if it's necessary.*

iconic design	Compare	well-loved	productively	version
intuition	first	Theme	Course	methodologies

1）Today I'm going to unpack for you three examples of _____.

2）It is a design that challenges your _____.

3）It is not what you think, it is when you _____ see it.

4）Here is my _____ iPhone.

5）Unlike the _____ Building, this is not alien technology.

6）Welcome to this Auto CAD Essential Training _____.

7）The whole idea of this course is that it'll teach you all the _____ and naming philosophies.

8）How to make sure that you drive Auto CAD _____ and most importantly profitably.

9）The first thing you should know is that we are using the latest _____ of Auto CAD.

10）Drawing _____ is a new feature in Auto CAD 2019.

3. *Work in groups: you are going to conduct a survey among your college peers on their favourite modern design. After your survey, prepare a PPT presentation of your survey result.*

4. *Translate the following sentences into English.*

1）这是一个设计，一旦你看到它，你就不会忘记它。

_____

_____

2）今天我将为大家展示三个标志设计的例子。

_____

_____

3）现在您可以从 web 和移动平台上打开，因此您可以从基于 web 的 Auto CAD 版本和基于移动平台的 Auto CAD 版本中打开。

_____

_____

4）在 Auto CAD 2019 中有一些很酷很有用的新功能。

_____

_____

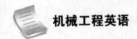

5）图纸比较是 Auto CAD 2019 中的一个新功能。

_____

_____

5. *Please write an e-mail about* 100 *words on the topic：**Inform your employees that you will be trained CAD courses.** Some specific examples will be highly appreciated and you have to watch out the spelling of some specialized terms you have learnt in this unit.*

_____

_____

_____

_____

_____

_____

_____

_____

_____

_____

VI. Additional Reading

The History of Autodesk, Inc.

Autodesk, Inc. is an American multinational software corporation that makes software for the architecture, engineering, construction, manufacturing, media, and entertainment industries. Autodesk is headquartered in San Rafael, California, and features a gallery（画廊）of its customers' work in its San Francisco building. The company has offices（办事处）worldwide, with U. S. locations in Northern California, Oregon（俄勒冈州）, Colorado, Texas, Michigan （密歇根州）and in New England in New Hampshire（新罕布什尔州）and Massachusetts, and Canada locations in Ontario, Quebec, and Alberta.

The company was founded in 1982 by John Walker, a coauthor（合著者）of the first versions of Auto CAD, the company's flagship（旗舰）computer-aided design（CAD）software. Its Auto CAD and Revit software are primarily（主要）used by architects, engineers, and structural designers to design, draft, and model buildings and other structures. Autodesk software has been used in many fields, and on projects from the One World Trade Center to Tesla（特斯拉）electric cars.

Autodesk became best known for Auto CAD, but now develops a broad range of software for design, engineering, and entertainment—and a line of software for consumers, including Sketchbook. The company makes educational versions of its software available at no cost to qualified students and faculty through the Autodesk Education Community, and also as a donation to eligible nonprofits through TechSoup Global. The manufacturing industry uses Autodesk's digital prototyping(数字原型制作)software—including Autodesk Inventor, Fusion 360, and the Autodesk Product Design Suite—to visualize, simulate, and analyze real-world performance using a digital model in the design process. The company's Revit line of software for building information modeling is designed to let users explore the planning, construction, and management of a building virtually before it is built.

Autodesk's Media and Entertainment division creates software for visual effects, color grading(分级), and editing as well as animation, game development, and design visualization. 3ds Max and Maya are both 3D animation software used in film visual effects and game development.

(*If you want to find more information about this corporation, please log on https://en. wikipedia. org/wiki/Autodesk.*)

1. *Read the passage quickly by using the skills of skimming and scanning, and choose the best answer to the following questions.*

    1) Which industry is not mentioned in the first paragraph? _____

        A. architecture   B. engineering

        C. manufacturing   D. computer

    2) Where is the headquarter of Autodesk? _____

        A. San Rafael   B. San Francisco

        C. Colorado   D. Texas

    3) The company was founded in _____ .

        A. 1980   B. 1981

        C. 1982   D. 1983

    4) Its Auto CAD and Revit software are not primarily used by architects, engineers, and structural designers to _____ .

        A. print   B. design

        C. draft   D. model buildings

    5) Autodesk software has been used in many fields, and on projects from the One World Trade Center to _____ electric cars.

        A. BMW   B. BYD

        C. Mini   D. Tesla

    6) The company makes _____ versions of its software available.

        A. cultural   B. educational

        C. latest   D. revised

    7) The manufacturing industry uses Autodesk's digital prototyping(数字原型制作) software—except _____ .

        A. digital prototyping   B. Autodesk Inventor

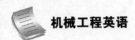

C. Fusion 360   D. Autodesk Product Design Suite

8) The company's Revit line of software for building information modeling is not designed to let users explore _____ before it is built.

A. design   B. planning

C. construction   D. management of a building virtually

9) What are the divisions of Autodesk? _____

A. Media   B. Entertainment

C. Media and Entertainment   D. Social

10) The division consists of _____ business units.

A. six   B. eight

C. nine   D. ten

2. *In this part, the students are required to make an oral presentation on either of the following topics.*

1) the history of Autodesk, Inc.

2) the lessons from Autodesk, Inc.'s development

习题答案

Unit Ten Engineering Materials

I. Pre-class Activity

Directions: *Please read the general introduction about* **Sprint** *and tell something more about the great scientist to your classmates.*

Applied materials company in the United States is the world's largest semiconductor(半导体) production equipment and technology service enterprise, as one of the fortune 500 global development growth enterprises, in 2001 it was named as one of the 100 most admired companies. The company started business in China as early as 1984 and became the first foreign semiconductor manufacturing equipment supplier in China. In 1997, the company was registered in China and its headquarter was located in Pudong, Shanghai.

Founded in 1967, Applied materials company is the world's largest nano-manufacturing(纳米制造) technology company and the largest supplier of equipment, services and software products in the electronics industry. Applied materials company has a history of more than 30 years in China and is the first foreign chip(芯片) manufacturing equipment company to enter the country. Currently, Applied materials company has become one of the top suppliers of semiconductor, flat-panel display(平板显示) and solar photovoltaic(光伏) manufacturing

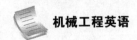

机械工程英语

equipment and services in China.

II. Specialized Terms

Directions：*Please memorize the following specialized terms before the class so that you will be able to better cope with the coming tasks.*

acrylic adj. 丙烯酸的

adjacent adj. 邻近的

advent n. 到来

blast vt. 爆炸;损害

cap vt. 覆盖;胜过

ceiling n. 天花板

coefficient n.系数

conductive adj. 传导的

consuming v. 消耗

curling n. 头发的卷曲;卷缩

dense adj. 稠密的

diffuse v. 散布

dig vt. 挖,掘

dug v. 挖,翻土

exterior n. 外部

fabric n. 织物

futuristic adj. 未来派的

glands n. 腺体

grid n. 网格;格子,栅格;输电网

hacker n. 电脑黑客

handset n. 手机,电话听筒

hard disk 硬盘

hard drive 硬盘驱动器

highly cost-effective 高性价比的

high-performance 高性能的;高效能的

high-volume 大容量的

hit the market 上市; 冲击市场, 打入市场

hybrid system 混合系统

hydraulic adj. 液压的

hydraulic drive 液压传动

hydraulic engineering 水利工程

immense adj. 极大的,巨大的

impurity n.［复数 impurities］杂质;不纯;
不洁

IMT abbr. 国际移动通信（International Mobile Telecommunications）; 智能多模式终端(Intelligent Multimode Terminal)

in as much as 因为,由于

in reference to 关于

in the first line 在第一线

indefinitely adv. 不确定的

industrial robot 工业机器人

information security 信息安全

infuse vt. 灌输

ink n. 墨水

innumerable adj. 无数的

inspection n. 检验

inspire vt. 激励;鼓舞

installed adj. 安装的

insulation n. 绝缘

insulator n.［物］绝缘体

intake n. 摄取量;通风口

integrated circuit 集成电路

integrating v. 整合

integration n. 整体化;集成;综合

intelligent network［计］［通信］智能网

intense adj. 强烈的;紧张的

interdisciplinary n. 各学科之间的

interior adj. 内部的

irreversible adj. 不可逆的

lamination v. 分层而成

manufacture n. 制造;产品

mass-produced 大(批)量生产的
material handling 物料输送
match plate 双面模版
match sand 假型砂
match wheel 配合轮
mating gear 配对齿轮
mechanics of material 材料力学
mechanism n. 力学
mechanization n. 机械化,机理
mechanized welding 机械化焊接
mentioned v. 提到
merge n. 合并;归并 v. 合并;融合
mesa transistor 台面晶体管
metal diaphragm 金属膜片
metal electrodeposition 金属电沉积
metal leaf 金属箔
metallic coating 金属涂层
metallic thermocouple 金属热电偶
metallurgical engineering 冶金工程
method n. 方法;条理;类函数
mistaken adj. 错误的;弄错的
modeling n. 建模;造型

motion n. 动作
multi-touch n. 多点触控
prosthetic adj. 假体的
organ n. 器官;机构
palette n. 调色板;颜料
particle n. 颗粒
pigment n. 色素
plastic n. 塑料制品
prototype n. 原型
reliance n. 信赖
rut n. 惯例
stuffy adj. 闷热的
textile n. 纺织品
thermo-bimetal n. 热双金属材料
user-friendly adj. 容易使用的
vacancy n. 空缺;空位;空白
vacuum n. 真空;空间
vacuum tube 真空管;电子管
valence n.[化学]价;[化学]原子价;[化学]化合价;效价
wire n. 电线

III. Watching and Listening

Task One How to Play the Intelligence Material

New Words

component n.成分;组件
textile n. 纺织品
futuristic adj. 未来派的
composite n. 复合材料
plastic n. 塑料制品
pigment n.[物]色素
fabric n. 织物
conductive adj. 传导的
ink n. 墨水

circuit n. 电路
wire n. 电线
acrylic adj. 丙烯酸的
infuse vt. 灌输
dense adj. 稠密的
diffuse v. 散布
particle n. 颗粒
interior adj. 内部的
multi-touch 多点触控

视频链接及文本

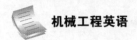

Exercises

1. *Watch the video for the first time and choose the best answer to the following questions.*

  1) Many of these do-it-yourself practices were lost in the second half of the _____ century.

    A. 17th   B. 18th

    C. 19th   D. 20th

  2. Where does her friend live? _____

    A. Portugal   B. Germany

    C. Spain   D. France

  3) What traditional materials are we talking about here? _____

    A. paper and textiles   B. paper

    C. textiles   D. stone

  4) Two of the known applications for this material include _____ design and multi-touch systems.

    A. modern   B. internal

    C. interior   D. external

  5) One of the principle applications for this materialis, amongst other things, in baby bottles, so it indicates when the contents are _____ enough to drink.

    A. warm   B. cool

    C. cold   D. hot

2. *Watch the video again and decide whether the following statements are true or false.*

  1) He did it because he could afford a car, but also because he knew how to build one. (   )

  2) Many of these do-it-yourself practices were lost in the second half of the 20th century. (   )

  3) For the most part, we still know what traditional materials like paper and textiles are made of and how they are produced. (   )

  4) They will be in many of the objects and technologies we use on a daily basis. (   )

  5) So conductive ink allows us to paint circuits instead of using the traditional printed circuit boards or wires. (   )

3. *Watch the video for the third time and fill in the following blanks.*

    I have a friend in _____ whose grandfather built a _____ out of a _____ and a washing machine so he could _____ his family. He did it because he couldn't _____ a car, but also because he knew how to _____ one. There was a time when we understood how things worked and how they were made, so we could build and _____ them, or at the very least make informed decisions about what to buy.

    Many of these _____ practices were lost in the second half of the 20th century. But now, the maker _____ and the open-source model are bringing this kind of knowledge about how things work and what they're made of back into our lives, and I believe we need to take them to the next level, to the _____ things are made of.

4. *Share your opinions with your partners on the following topics for discussion.*

1) Do you like the lecture about intelligence material? Why do you enjoy such a video?

2) Can you use a few lines to list what's your understanding about intelligence material? Please use an example to clarify your thoughts.

Task Two　Metal that Breathes

New Words

视频链接及文本

stuffy adj. 闷热的
cap vt. 覆盖;胜过
consuming v. 消耗
blast vt. 爆炸;损害
insulation n. 绝缘
exterior n. 外部
advent n. 到来
ceiling n. 天花板
irreversible adj. 不可逆的;多样
reliance n. 信赖
mechanical adj. 机械的
massive adj. 大量的
adjacent adj. 邻近的

rut n. 惯例
dig vt. 挖,掘
dug v. 挖,翻土
organ n. 器官;机构
glands n. 腺体
palette n. 调色板;颜料
thermo-bimetal n. 热双金属材料
lamination v. 分层而成
coefficient n. [数]系数
curling n. 头发的卷曲;卷缩;冰上溜石
　　游戏
prototype n. 原型
ventilate vt. 使通风

Exercises

1. *Watch the video for the first time and choose the best answer to the following questions.*

1) My father would not let us use the air conditioner? _____
A. cold　　　　　　　　　　B. overheat
C. overheat the engine　　　　D. heat

2) What is the feature of thick walls? _____
A. cool in summer　　　　　　B. temperature transfer
C. warm in winter　　　　　　D. both of all

3) Then in about the _____, with the advent of plate glass.
A. 1910s　　　　　　　　　　B. 1920s
C. 1940s　　　　　　　　　　D. 1930s

4) I propose that our building skin should be more similar to _____ skin.
A. human　　　　　　　　　　B. animal
C. color　　　　　　　　　　D. shape

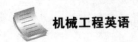

5）Keep in mind, with the digital technology that we have today, this thing was made out of about _____ pieces and there's no two pieces alike at all. Every single one is different.

    A. 4,000   B. 24,000

    C. 14,000   D. 34,000

2. *Watch the video again and decide whether the following statements are true or false.*

1）We don't have cars that we take across country. (　　)

2）We blast the air conditioning the entire way, and we never experience overheating. (　　)

3）In the past, before air conditioning, we didn't have thick walls. (　　)

4）Over time, the buildings got taller and bigger, our engineering even better, so that the mechanical systems were massive. (　　)

5）I propose that our building skin should be more similar to human skin, and by doing so it can be much more dynamic, responsive and different, depending on where it is. (　　)

3. *Watch the video for the third time and write at least three materials and application fields of building.*

Materials	Building

4. *Share your opinions with your partners on the following topics for discussion.*

1）What's your impression of these materials?

2）discussion of the operation and application of materials in construction

IV. Talking

Task One　Classical Sentences

Directions: *In this section, some popular sentences are supplied for you to read and to memorize. Then, you are required to simulate and produce your own sentences with reference to the structure.*

General Sentences

1. What channel did you watch on television last night?

    昨天晚上你看的哪个频道的电视节目？

2. I don't get a good picture on my TV set. There's probably something wrong.
我的电视机画面不清晰,可能出毛病了。

3. You get good reception on your radio.
你的收音机接收效果很好。

4. Please turn the radio up. It's too low.
请把收音机开大点声,声音太小了。

5. What's on following the news and weather?
新闻和天气预报后是什么节目?

6. Do you have a TV guide?
你有电视节目指南吗?

7. You ought to have Bill look at your TV. Maybe he could fix it.
你应该让比尔检查下你的电视,他可能会修理。

8. We met one of the engineers over at the television station.
我们在电视台遇见了一个在那里工作的工程师。

9. Where can I plug in the TV? Is this outlet all right?
电视插头该插在哪里? 这个插头可以用吗?

10. I couldn't hear the program because there was too much static.
因为干扰太大,我听不清节目了。

11. Your car radio works very well. What kind is it?
你的车载收音机性能很好,它是什么类型的?

12. The next time I buy a TV set, I'm going to buy a portable model.
下一次我买电视机时,我打算买一部便携式的。

13. I wonder if this is a local broadcast.
我想知道这是不是本地广播。

14. You'd get better TV reception if you had an outside antenna.
如果你有外部天线的话,电视机的接收效果会更好。

15. Most amateur radio operators build their own equipment.
大部分业余收音机爱好者都自己装收音设备。

16. Station Voice of WIT is off the air now. They signed off two hours ago.
WIT 电台已经停止广播了,它们在两小时之前就结束了。

17. Who is the author of this novel?
这部小说的作者是谁?

18. I've never read a more stirring story.
我从没有读过这么感人的故事。

19. Who would you name as the greatest poet of our time?
你认为谁是我们这个时代最伟大的诗人?

20. This poetry is realistic. I don't care for it very much.
这本诗集是写实的,我不太喜欢。

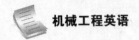

21. Many great writers were not appreciated fully while they were alive.

许多大作家在世时并没有得到人们的赏识。

22. This is a poem about frontier life in the United States.

这是一首描述美国边境生活的诗。

23. This writer uses vivid descriptions in his writings.

这位作者作品中的描述非常生动。

24. How much do you know about the works of Henry Wadsworth Longfellow?

关于亨利·沃兹沃思·朗费罗的著作,你了解多少?

25. As the saying goes, where there is a will, there is a way.

俗话说,有志者事竟成。

26. It is well-known to all that all roads lead to Rome.

众所周知,条条大路通罗马。

27. Whatever is worth doing is worth doing well.

任何值得做的事都值得做好。

28. The hardest thing to learn is to be a good loser.

最难学的是做一个输得起的人。

29. Happiness is a way station between too much and too little.

幸福,是位于太多和太少之间的一个小站。

30. The hard part isn't making the decision. It's living with it.

做出决定并不困难,困难的是接受这个决定。

31. You may be out of my sight, but never out of my mind.

你也许已走出我的视线,但从未走出我的思念。

32. Love is not a maybe thing. You know when you love someone.

爱不是什么可能、大概、也许,一旦爱上了,自己是十分确定的。

33. In the end, it's not the years in your life that count. It's the life in your years.

到头来,你活了多少岁不算什么,重要的是,你是如何度过这些岁月的。

34. When the whole world is about to rain, let's make it clear in our heart together.

当全世界约好一起下雨时,我们约好一起在心里放晴。

35. It's better to be alone than to be with someone you're not happy to be with.

宁愿一个人待着,也不要跟不合拍的人待一块。

36. One needs three things to be truly happy living in the world: something to do, someone to love, and something to hope for.

在这个世界上,我们只需拥有三样东西便可真正快乐:有事可做,有人去爱,充满希望。

37. No matter how badly your heart has been broken, the world doesn't stop for your grief. The sun comes right back up the next day.

不管你有多痛苦,这个世界都不会因你悲伤而停止转动,太阳第二天依旧升起。

38. Accept what was and what is, and you'll have more positive energy to pursue what

will be.

接受过去和现在的模样,你才有能量去追寻自己的未来。

39. Until you make peace with who you are, you'll never be content with what you have.

除非你能和真实的自己和平相处,否则你永远不会对已拥有的东西感到满足。

40. If you wish to succeed, you should use persistence as your good friend, experience as your reference, prudence as your brother and hope as your sentry.

如果你希望成功,当以恒心为良友,以经验为参谋,以谨慎为兄弟,以希望为哨兵。

41. Great minds have purpose, others have wishes.

杰出的人有着目标,其他人只有愿望。

42. Being single is better than being in an unfaithful relationship.

比起谈着充满欺骗的恋爱,单身反而更好。

43. If you find a path with no obstacles, it probably doesn't lead anywhere.

太容易的路可能根本就不会带你去任何地方。

44. Getting out of bed in winter is one of life's hardest mission.

冬天起床是人生最艰难的任务之一。

45. The future is scary but you can't just run to the past because it's familiar.

未来会让人心生畏惧,但是我们却不能因为习惯了过去就逃回过去。

46. Success is the ability to go from one failure to another with no loss of enthusiasm.

成功是,你即使跨过一个又一个失败,也没有失去热情。

47. Not everything that is faced can be changed, but nothing can be changed until it is faced.

并不是你面对了,事情就能改变。但是,如果你不肯面对,那什么也改变不了。

48. If they throw stones at you, don't throw back. Use them to build your own foundation instead.

如果别人朝你扔石头,不要扔回去了,留着做你建高楼的基石吧!

49. If your happiness depends on what somebody else does, I guess you do have a problem.

如果你的快乐与否取决于别人做了什么,我想,你真的有点问题。

50. Today, give a stranger one of your smiles. It might be the only sunshine he sees all day.

今天,给一个陌生人送上你的微笑吧,很可能,这是他一天中见到的唯一的阳光。

Specialized Sentences

1. Solid materials have been conveniently grouped into three basic classifications: metals, ceramics, and polymers.

固体材料被简单地分为三个基本类型:金属、陶瓷和聚合物。

2. This scheme is based primarily on chemical makeup and atomic structure, and most materials fall into one distinct grouping or another, although there are some intermediates.

该方案主要是基于化学组成和原子结构,尽管存在一些中间产物,大多数材料被分为一个不同的组或另一组。

3. There are three other groups of important engineering materials—composites, semiconductors,

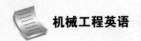

and biomaterials.

有三类其他重要的工程材料——复合材料、半导体材料和生物材料。

4. Composites consist of combinations of two or more different materials, whereas semiconductors are utilized because of their unusual electrical characteristics.

复合材料由两种或者两种以上不同的材料组成,而半导体材料因其非同寻常的电学性质而得到使用。

5. Biomaterials are implanted into the human body.

生物材料被移植到人类的身体中。

6. A brief explanation of the material types and representative characteristics is offered next.

关于材料类型及其特征的一个简单的解释将在后面给出。

7. Metallic materials are normally combinations of metallic elements.

金属材料通常由金属元素组成。

8. They have large numbers of non-localized electrons.

它们有大量无规则运动的电子。

9. Many properties of metals are directly able to these electrons.

金属的许多性质直接作用于这些电子。

10. Metals are extremely good conductors of electricity and heat, and are not transparent to visible light.

金属是电和热的极好导体,它们不透光。

11. A polished metal surface has a lustrous appearance.

一个抛光的金属表面有光泽。

12. Furthermore, metals are quite strong, yet deformable, which accounts for their extensive use in structural applications.

除此之外,金属十分坚硬,但可变形,这是它们在结构方面使用广泛的原因。

13. Ceramics are compounds between metallic and nonmetallic elements.

陶瓷是介于金属元素和非金属元素之间的化合物。

14. So these are just a few of what are commonly known as smart materials.

这些材料通常被称为智能材料。

15. The wide range of materials that falls within this classification includes ceramics that are composed of clay minerals, cement, and glass.

这类材料的范围很广,包括由黏土矿物、水泥和玻璃组成的陶瓷。

16. These materials are typically insulative to the passage of electricity and heat, and are more resistant to high temperatures and harsh environments than metals and polymers.

这些材料是典型的电和热的绝缘体,而且它们比金属和聚合物更加适于高温和严苛的环境。

17. With regard to mechanical behavior, ceramics are hard but very brittle.

至于机械性能,陶瓷硬但很脆。

18. Polymers include the familiar plastic and rubber materials.

聚合物包括常见的塑料和橡胶材料。

19. Many of them are organic compounds that are chemically based on carbon, hydrogen, and other nonmetallic elements.

它们中的大多数是有机化合物,这些化合物是用化学的方法把碳、氢和其他非金属元素组合而成的。

20. These materials typically have low densities and may be extremely flexible.

这些材料通常密度低,并且十分柔软。

21. A number of composite materials have been engineered that consist of more than one material type.

许多复合材料被用于工程中,它们由至少一种类型以上的材料组成。

22. Fiberglass is a familiar example, in which glass fibers are embedded within a polymeric material.

玻璃丝是一个常见的例子,其中玻璃纤维被嵌入高分子材料中。

23. A composite is designed to display a combination of the best characteristics of each of the component materials.

一种复合材料被设计出来是为了联合显示每一种组分材料的最佳特性。

24. Fiberglass acquires strength from the glass and flexibility from the polymer.

玻璃丝从玻璃中获得强度,从聚合物中获得柔韧性。

25. Many of the recent material developments have involved composite materials.

最近绝大多数材料的发展都涉及复合材料。

26. Semiconductors have electrical properties that are intermediate between the electrical conductors and insulators.

半导体有电的性质,是介于电导体和绝缘体之间的中间物。

27. The electrical characteristics of these materials are extremely sensitive to the presence of minute concentrations of impurity atoms, which concentrations may be controlled over very small spatial regions.

这些材料的电学性质对微量杂质原子的存在十分敏感,杂质原子浓度可能控制在一个十分小的区域内。

28. Biomaterials are employed in components implanted into the human body for replacement of diseased or damaged body parts.

生物材料被移植进入人类身体,以取代病变的或者损坏的身体部件。

29. These materials must not produce toxic substances and must be compatible with body tissues.

这些材料不能产生有毒物质,而且必须同人的身体器官兼容。

30. Materials that are utilized in high-technology (or high-tech) applications are sometimes termed advanced materials.

用在高科技中的材料有时被称作高级材料。

31. By high technology we mean a device or product that operates or functions using relatively

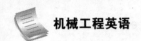

intricate and sophisticated principles.

所谓高科技是指使用相对复杂和熟练的原理来操作或运行的设备。

32. Examples include electronic equipment , computers, fiberoptic systems, spacecraft, aircraft, and military rocketry.

这些例子包括电子设备、计算机、光纤系统、宇宙飞船、飞机和军事火箭。

33. In spite of the tremendous progress that has been made in the discipline of materials science and engineering within the past few years.

尽管在过去几年内,材料科学与工程领域取得了巨大的进步。

34. There still remain technological challenges, including the development of even more sophisticated and specialized materials, as well as consideration of the environmental impact of materials production.

仍然存在一些技术挑战,包括开发更加复杂和专业的材料,并且要考虑材料生产对环境产生的影响。

35. Nuclear energy holds some promise, but the solutions to many problems that remain will necessarily involve materials, from fuels to containment structures to facilities for the disposal of radioactive waste.

核能仍有一定前景,但是要解决许多的问题,涉及从燃料到安全结构,再到放射性废物处理设施的各种材料。

36. Significant quantities of energy are involved in transportation.

相当数量的能源用在交通上。

37. New high strength, low-density structural materials remain to be developed, as well as materials that have higher-temperature capabilities, for use in engine components.

新的高强度、低密度结构材料仍在发展中,用作引擎部位的耐高温材料也在发展中。

38. There is a recognized need to find new, economical sources of energy, and to use the present resources more efficiently.

寻找新的、经济的能源资源,并且更加有效地使用目前现存的资源是公认为必需的。

39. Materials will undoubtedly play a significant role in these developments.

材料将毫无疑问地在这些发展中扮演重要的角色。

40. The direct conversion of solar into electrical energy has been demonstrated.

太阳能直接转化为电能的方法已经被证实。

41. Solar cells employ some rather complex and expensive materials.

太阳能电池使用了一些相当复杂且昂贵的材料。

42. To ensure a viable technology, materials that are highly efficient in this conversion process yet less costly must be developed.

为了保证技术的可行性,必须开发在这个转化过程中效率高、成本低的材料。

43. Environmental quality depends on our ability to control air and water pollution.

环境质量取决于我们控制大气和水污染的能力。

198

44. Pollution control techniques employ various materials.
污染控制技术使用了各种材料。

45. Less pollution and less despoilage of the landscape from the mining of raw materials.
采矿过程中更少的污染和对环境更少的破坏。

46. Also, in some materials manufacturing processes, pay attention to environmental protection.
另外,在一些材料的生产过程中要注意环保。

47. Many materials that we use are derived from resources that are nonrenewable, that is, not capable of being regenerated.
我们使用的许多材料来源于不可再生的资源,不可再生即不能再次生成的。

48. These include polymers, for which the prime raw material is oil, and some metals.
这些材料包括聚合物,其最初的原材料是油和一些金属。

49. These nonrenewable resources are gradually becoming depleted.
这些不可再生的资源逐渐变得枯竭。

50. Toxic substances are produced, and the ecological impact of their disposal must be considered.
有毒物质产生了,并且它们的处置对生态产生的影响必须加以考虑。

Task Two   Sample Dialogue

Directions: *In this dialogue, you are going to read several times the following sample dialogue about the relevant topic. Please pay special attention to five Cs (culture, context, coherence, cohesion and critique) in the dialogue and get ready for a smooth communication in the coming task.*

Design of elevators

(*George and Steven are talking about the design of an elevator.*)

George: Hello, this is George White. Speaking, please.

Steven: Good afternoon, Mr. White. This is Steven Baker, a designer of Cascade Elevator Company.

George: Oh, Mr. Baker. I'm looking forward to your phone call. I went to your company the other day, but you were not there.

Steven: I was on a business trip. Our general manager Mr. Cox told me that I would be in charge of the design of your elevator system.

George: He mentioned that to me. I know you are very experienced.

Steven: Thank you, Mr. White. Actually, we'll have a team of five designers working on this project.

George: Great. Did you get the information about our building and our requirement for the design?

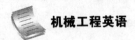

Steven: Yes. I've gone through your papers. And we'll have the first meeting on the project tomorrow morning. It would be great if you can come to give a detailed illustration.

George: OK. What is the time?

Steven: 10 o'clock.

George: No problem. I'll be there. Bye.

Steven: Bye.

Task Three　Simulation and Reproduction

Directions: *The class will be divided into three major groups, each of which will be assigned a topic. In each group, some students may be the teacher, while others may be students. In the process of discussion, please observe the principles of cooperation, politeness and choice of words. One of the groups will be chosen to demonstrate the discussion to the class.*

1) materials in our daily life

2) the application of materials

3) the importance of learning engineering materials

Task Four　Discussion and Debate

Directions: *The class will be divided into two groups. Please choose your stand in regard to the following controversy and support your opinions with scientific evidences. Please refer to the specialized terms and classical sentences in the previous parts of this unit.*

Materials may be grouped in several ways. What kinds of engineering materials are divided into? Conld you explain it?

V. After-class Exercises

1. *Match the English words in Column A with the Chinese meaning in Column B.*

A	B
1) dug	A) 挖,翻土
2) glands	B) 器官;机构
3) palette	C) 腺体
4) thermo-bimetal	D) 调色板;颜料
5) lamination	E) 热双金属材料
6) prototype	F) 分层而成
7) curling	G) [数]系数
8) coefficient	H) 冰上溜石游戏
9) organ	I) 原型

Unit Ten Engineering Materials

10) mechanical   J) 机械的

2. *Fill in the following blanks with the words or phrases in the word bank. Change the forms if it's necessary.*

practices	ink	textiles	interior	blast
massive	palette	roll	thermo-bimetal	Portugal

1) I have a friend in _____ whose grandfather built a vehicle out of a bicycle and a washing machine so he could transport his family.

2) Many of these do-it-yourself _____ were lost in the second half of the 20th century.

3) We still know what traditional materials like paper and _____ are made of and how they are produced.

4) So conductive _____ allows us to paint circuits instead of using the traditional printed circuit boards or wires.

5) Two of the known applications for this material include _____ design and multi-touch systems.

6) I was one of those kids that, every time I got in the car, I basically had to _____ down the window.

7) We _____ the air conditioning the entire way, and we never experience overheating.

8) Over time, the buildings got taller and bigger, our engineering even better, so that the mechanical systems were _____.

9) What I proposed first doing is looking at a different material _____ to do that.

10) I presently, or currently, work with smart materials, and a smart _____.

3. *Can you name ten materials?*

   1)_____ 2)_____ 3)_____ 4)_____ 5)_____
   6)_____ 7)_____ 8)_____ 9)_____ 10)_____

4. *Translate the following sentences into English.*

   1) 它被称为"Bloom",它的表面完全由热双金属制成,其目的是让这个树冠能做两件事。

   _____

   2) 第一,它是一个遮阳装置,当太阳照射到地面时,它会限制阳光通过的量,另外,它是一个通风系统。

   _____

   3) 所以,下面被困的热空气可以在必要的时候进出。

   _____

   4) 这种材料的一个主要应用是在婴儿奶瓶中,它指示了什么时候里面的东西够凉可

201

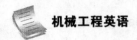

机械工程英语

以喝。

_____

5) 这些材料通常被称为智能材料。

_____

5. *Please write a letter on the subject:* ***How to write an application letter.*** *Some specific examples will be highly appreciated and watch out the spelling of some specialized terms you have learnt in this unit.*

_____
_____
_____
_____
_____
_____
_____
_____
_____
_____
_____

VI. Additional Reading

Applied Materials, Inc.

Applied Materials, Inc. is an American corporation that supplies equipment, services and software to enable the manufacture of semiconductor(半导体) integrated circuit(集成电路) chips(芯片) for electronics, flat panel displays(平板显示器) for computers, smartphones and televisions, and solar(太阳能) products. The company also supplies equipment to produce coatings for flexible electronics, packaging and other applications. The company is headquartered in Santa Clara, California, in Silicon Valley(硅谷).

Founded in 1967 by Michael A. McNeilly and others, Applied Materials went public(上市) in 1972. In subsequent years, the company was diversified, until James C. Morgan became CEO in 1976 and returned the company's focus to its core business(核心业务) of semiconductor(半导体) manufacturing equipment. By 1978, sales increased by 17%.

202

In 1984, Applied Materials became the first U. S. semiconductor equipment manufacturer to open its own technology center in Japan and the first semiconductor equipment company to operate a service center in China. In 1987, Applied Materials introduced a CVD machine called the Precision 5000, which differed from existing machines by incorporating diverse processes into a single machine that had multiple process chambers(处理室).

In 1992, the corporation settled a lawsuit(诉讼) with three former employees for an estimated $600,000. The suit complained that the employees were driven out of the company after complaining about the courses Applied Scholastics had been hired to teach there.

In 1993, the Applied Materials' Precision 5000 was inducted into the Smithsonian Institution's permanent collection(系列) of Information Age technology.

In November 1996, Applied Materials acquired two Israeli(以色列) companies for an aggregate amount of $285 million. Opal Technologies and Orbot Instruments for $175 million and $110 million in cash, respectively. Orbot produces systems for inspecting patterned silicon wafers(图案化硅晶片的系统) for yield enhancement during the semiconductor manufacturing process, as well as systems for inspecting masks used during the patterning process. Opal develops and manufactures high-speed metrology systems(高速计量系统) used by semiconductor manufacturers to verify critical dimensions during the production of integrated circuits(集成电路).

In 2000, Etec Systems, Inc. was purchased(被收购).

On June 27, 2001, Applied Materials acquired(收购) Israeli company Oramir Semiconductor Equipment Ltd. , a supplier of laser cleaning technologies for semiconductor wafers, in a purchase business combination for $21 million in cash.

In January 2008, Applied Materials purchased an Italian company Baccini, a designer of tools used in manufacturing solar cells.

In 2009, Applied Materials opened its Solar Technology Center—the world's largest commercial solar energy research and development facility in Xi'an, China.

Applied Materials' acquisition of Semitool Inc. was completed in December 2009.

Applied Materials announced its acquisition of Varian Semiconductor in May 2011.

Applied Materials announced its merger with Tokyo Electron on September 24, 2013. If approved by government regulators, the combined company, to be called Eteris, would be the world's largest supplier of semiconductor processing equipment, with a total market value of more than $30 billion.

But on April 27, 2015, Applied Materials announced that its merger with Tokyo Electron has been scrapped due to fears of dominating the semiconductor equipment industry.

Applied Materials is named among FORTUNE World's Most Admired Companies in 2018.

(*If you want to find more information about this corporation, please log on* https://en. m. wikipedia. org/wiki/Applied_Materials.)

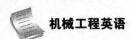

1. *Read the passage quickly by using the skills of skimming and scanning, and choose the best answer to the following questions.*

1) Where is the company's headquarter in California? _____

    A. Geneva   B. Basel

    C. Lyss   D. Santa Clara

2) When was the company founded? _____

    A. 1972   B. 1976

    C. 1967   D. 1978

3) In 1984, Applied Materials became the first U.S. semiconductor equipment manufacturer to open its own technology center in _____.

    A. America   B. UK

    C. U.S.   D. Japan

4) In November 1996, Applied Materials acquired two Israeli(以色列) companies for an aggregate amount of _____ million.

    A. $285   B. $175

    C. $110   D. $185

5) In January 2008, Applied Materials purchased an _____ company Baccini, a designer of tools used in manufacturing solar cells.

    A. Germany   B. Italian

    C. Chinese   D. Japanese

6) In 2009, Applied Materials opened its Solar Technology Center—the world's largest commercial solar energy research and development facility in _____, China.

    A. Beijing   B. Xi'an

    C. Shanghai   D. Guangzhou

7) The world's largest supplier of semiconductor processing equipment, with a total market value of more than _____.

    A. $4.2 billion   B. $30 million

    C. $30 billion   D. $3.9 billion

8) When does Applied Materials completed its merger of Tokyo Electron? _____

    A. June 30, 2018   B. September 24, 2013

    C. January 30, 2012   D. June 15, 2012

9) Applied Materials is named among FORTUNE world's most admired companies in _____.

    A. 2018   B. 2017

    C. 2016   D. 2015

10) Applied Materials' acquisition of Semitool Inc. was completed in _____ 2009.

    A. December   B. November

    C. October   D. September

2. *In this part, the students are required to make an oral presentation on either of the following topics.*

　1) the history of Applied Materials

　2) the lessons from Applied Materials' development history

习题答案